SALAM + 50

Proceedings of the Conference

SALAM + 50

Proceedings of the Conference

edited by

Michael Duff

Imperial College London, UK

ICP

Imperial College Press

Published by

Imperial College Press
57 Shelton Street
Covent Garden
London WC2H 9HE

Distributed by

World Scientific Publishing Co. Pte. Ltd.
5 Toh Tuck Link, Singapore 596224
USA office: 27 Warren Street, Suite 401-402, Hackensack, NJ 07601
UK office: 57 Shelton Street, Covent Garden, London WC2H 9HE

British Library Cataloguing-in-Publication Data
A catalogue record for this book is available from the British Library.

SALAM + 50
Proceedings of the Conference

ISBN-13 978-1-84816-190-0
ISBN-10 1-84816-190-5

Printed in Singapore by World Scientific Printers

CONTENTS

PREFACE

The year 2007 marked not only the centenary of Imperial College London but also the 50th anniversary of the late Abdus Salam's arrival at the College. Accordingly, a conference entitled "Salam+50" and organized by the Theoretical Physics Group, was held at Imperial College on July 7, 2007.

We were fortunate that many distinguished guests were able to attend and pay their respects to the great man. Their contributions recorded in these Proceedings are divided into three sections, designed to reflect the three dimensions of Salam's character: I. Salam the Scientist, II. Salam the Humanitarian and III. Salam the Man.

Following the welcoming address by the Rector of Imperial College, Sir Richard Sykes, Section I begins with recollections of Salam's contributions to Physics by Nobel Laureate Gerard 't Hooft, and Salam's former colleagues at Imperial Tom Kibble and Chris Isham (whose contribution formed the after-dinner speech). Also at the conference Kelly Stelle read out an English translation of an obituary of Salam written in Arabic by the late Yuval Ne-eman shortly before his own death. He had specifically requested Kelly that it be read at any future celebration of Salam at Imperial.

Section II includes discussions of Salam's involvement with Pakistan by Dr. Maleeha Lodhi, the UK Pakistani High Commissioner, and with the International Centre for Theoretical Physics (ICTP) by Faheem Hussain, formerly its Head of External Relations. The Director of the ICTP, K. R. Sreenivasan, was unfortunately unable to travel to London but kindly provided his personal account of Salam which is also included in this section.

No one knew Salam better than his family and we were delighted that many of them were able to attend and that three of them were able to speak: his son Ahmad Salam, his daughter Aziza Rahman and his grandson Saif Rahman. Their recollections are presented in Section III along with those of Gordon Fraser whose biography of Salam is shortly to be published. Salam's spiritual life is recorded by Mirza Masroor Ahmad, Head of the Worldwide Ahmadiyya Muslim Community. Finally there is the speech of Dr. Mirza Ahmad, President of the Ahmadiyya Muslim Youth Association UK, who kindly presented the Abdus Salam Imperial College Physics Graduate Prize to David Weir.

I would like to add my personal thanks to all the speakers for making the conference such a success, to Dr. Louise Johnson for her help and advice, to my fellow organizers Tom Kibble, Chris Isham, Kelly Stelle and Arttu Rajantie, to our secretary Graziela De Nadai-Sowrey, to Tariq Hayat for his assistance, and to Sir Peter Knight FRS, Principal of the Faculty of Natural Sciences, and Donal Bradley, Head of the Physics Department, for their support. Financial assistance from the Science and Technology Facilities Council is gratefully acknowledged.

Michael Duff
Abdus Salam Professor of Theoretical Physics
Imperial College London
2007

WELCOMING ADDRESS

SIR RICHARD SYKES, FRS
Rector, Imperial College London

Hello and thanks to everyone for being here this morning to celebrate the life and achievements of a remarkable man. It's safe to say that Abdus Salam was a one off in many ways, so this promises to be a fascinating and inspiring day. I'm especially delighted that we have with us members of the Salam family — it's an honour to welcome them to Imperial. It's a happy coincidence that Imperial marks its Centenary in the same year that it also marks the 50th anniversary of the arrival of one of its leading scientists.

Abdus Salam is someone who still has a great deal to teach us, and not just about physics. He was certainly a great scientist, a Nobel Laureate in a very complex field that I'm not even going to try to explain in front of a roomful of physicists. He came to Imperial in 1957, having previously studied and worked in both Lahore and Cambridge. I'm pleased to say that he stayed here for the rest of his career, and retained his connections with the College until his death in 1996.

He obviously showed great promise right from the start of his academic career, gaining the highest marks ever recorded in the Matriculation Examination at the University of the Punjab. He went on to set up the theoretical physics group here at Imperial, while at the same time keeping his links with his homeland strong. He had a keen interest in the scientific development of Pakistan and was a member of the Pakistan Atomic Energy Commission, a member of the Scientific Commission of Pakistan and acted as Chief Scientific Advisor to the President. When he was awarded the Nobel Prize in 1979, he had the distinction of becoming the first Muslim and the first Pakistani Nobel Laureate.

Abdus was known and admired for far more than his scientific discoveries. He firmly believed that "scientific thought is the common heritage of mankind", and did much to advance that conviction. In 1964 he founded the International Centre for Theoretical Physics in Trieste, Italy, where he set up associateships for gifted researchers from developing countries. He went on to found the Third World Academy of Sciences, seeing investment in the science base of developing countries as a key way to reduce the gap between rich and poor. In this way, he believed, science could play a vital role in promoting world peace.

I said earlier that Abdus was the first Muslim Nobel Laureate, and he remained committed to his faith throughout his life. We probably all know that religion and science haven't always been happy bed-fellows, but he saw no contradiction between the two. He quoted the Quran during his Nobel speech and wrote of his work as a physicist: "That our generation has been privileged to glimpse a part of God's design is a bounty and a grace for which I render thanks with a humble heart."

This is just a short introduction to a many-faceted man. I hope it has given a taste of why Abdus Salam is remembered with such respect and fondness by all who knew him. We have a programme of distinguished speakers here today who I'm sure will be able to give us a new insight into his work and wider interests. Thanks are due to Mike Duff and the other organisers for putting together this varied and stimulating programme. I hope that by coming together in this way we can do justice to the memory of a truly great man.

Salam the Scientist

THE GRAND VIEW OF PHYSICS

GERARD 't HOOFT
Institute for Theoretical Physics, Utrecht University
and
Spinoza Institute, Postbox 80.195, 3508 TD Utrecht, The Netherlands
g.thooft@phys.uu.nl
http://www.phys.uu.nl/~thooft/

Abdus Salam was known for his "grand views", grand views of science as well as grand views of society. In this talk the grand view of theoretical physics is put in perspective.

1. Confronting Challenges

To obtain the Grand Picture of the physical world we inhabit, to identify the real problems and distinguish them from technical details, to spot the very deeply hidden areas where there is room for genuine improvement and revolutionary progress, courage is required. Every now and then, one has to take a step backwards, one has to ask silly questions, one must question established wisdom, one must play with ideas like being a child. And one must not be afraid of making dumb mistakes.

By his adversaries, Abdus Salam was accused of all these things. He could be a child in his wonder about beauty and esthetics, and he could make mistakes. Glancing back at his numerous and wildly varied publications, I can see why some people found it difficult to understand why he was given the most prestigious award of our trade, the Nobel Prize, since even the one publication that is cited most, his work with John C. Ward, places the leptons in a multiplet that no one today would find acceptable, and even in his days, it could easily be argued why that proposition would have to fail. Salam himself was also surprised by the quotation that went with his Nobel Prize. "I think", he once confided to me, "that the Nobel Committee also rewarded me for my ideas about the two-component neutrino. That was right on the spot, and I was the first."

But that was not what he was rewarded for. I am sure that the Nobel Committee would have mentioned it. Instead, the committee stated:

> "for their contributions to the theory of the unified weak and electromagnetic interaction between elementary particles, including, inter alia, the prediction of the weak neutral current".

Indeed, the existence of a neutral component in the weak current happened to be correctly predicted even if you took the wrong multiplet. The real reason why he earned the Nobel was that he had the grand view. As Sidney Coleman had noticed, "Salam even gave the essence of a correct argument for his belief in the renormalizability of the theory". And, one cannot emphasize enough, *that* is what counts. After all, we now know that what is called the "Standard Model" today, is just one apparently haphazard choice, made by Nature, among many possibilities of principle. Salam's multiplet *could* have been right; his grand vision certainly was right.

This grand view evolved extensively. The principle that truly dominated the later half of the 20th century was the principle of symmetry. "If you can identify Nature's complete symmetry group, you will know everything", is what became a pivotal dogma. But before this insight was truly appreciated, one first had to overcome a major obstacle: divergences. Are the divergences in the integration expressions for quantized field theories a fundamental shortcoming of the general idea, or can they be overcome, so that the divergences can be viewed as nothing more than a temporary technical obstacle? Salam had been strongly attracted to this question. But indeed he posed the question, rather than airing dogmatic views on the subject, like so many of his contemporaries.

Did he see it right? Salam did notice that overlapping divergences can be disentangled, that divergences in many theories can indeed be seen as a technical problem that does not disqualify quantized field theories as a whole, and that exact gauge symmetry is essential for the theories to work. But then he played and made mistakes. Any attempt to quantize gravity is also beset by divergences. Well, having put aside the divergence problem as a technical one, perhaps it is a technical problem in gravity as well? When I met him frequently, he was fiercely attempting to rearrange the Feynman diagrams of quantum gravity. Since the interaction is non-polynomial (in most gauges, that is), one can observe that any pair of vertices in gravity could be connected by an indefinite number of propagators, the *superpropagator* (see Fig. 1).

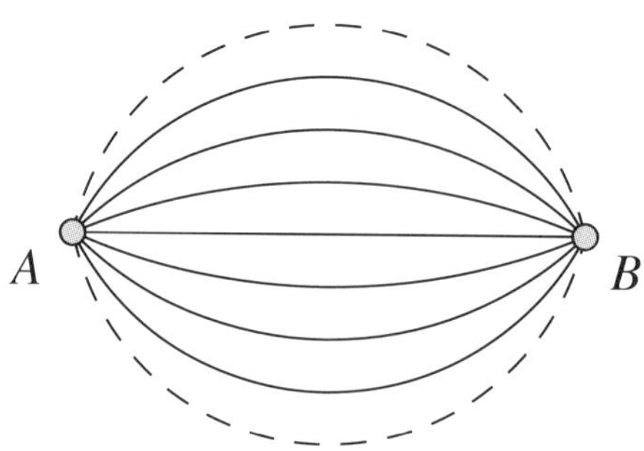

Fig. 1. The superpropagator. Two vertices, A and B, are connected by an indefinite number of elementary propagators.

As many other attempts to tame quantum gravity, this one was bound to fail. Here I never could agree with Salam, for very fundamental reasons. My problem with this is the same difficulty I have with claims from both supergravity and string theory today. Just imagine that indeed the integrals that one needs to perform would converge, or at least become sufficiently regular, whenever the momentum variables k_μ would tend to infinity. This would mean that some kind of decoupling would take place at very high momentum, or equivalently, at very tiny distance scales. This decoupling would indicate that, at tiny distance scales, our system would linearize, decouple, simplify. We would be able to describe a smooth and comprehensible world at distance scales smaller than, say, the Planck length. Why not try to imagine such a world directly? The point is that this is impossible. Newton's constant would tend to infinity there, or, distances in space and time, as what we are familiar with, cease to make sense there. This is characteristic for a *topological* theory. Thus, gravity must become purely topological at small distances. As long as we do not have such a topological theory, chances that we stumble upon one by blindly manipulating superpropagators, supergravity diagrams or string world sheets, are remote. Our searches should be well directed ones.

2. Grand Unification

Should the color group SU(3) be extended to SU(4) so as to "unify" the leptons with the quarks? This is what Salam thought, and he further pursued the thought with Jogesh Pati, ideas that were also expanded by Georgi, Glashow, Quinn and De Rujula. It so turned out that the fermions most naturally fit in a $10 \oplus \bar{5}$ representation of SU(5), a group that combined color SU(3) with electroweak SU(2) $\times$ U(1). It later turned out that this SU(5) did not quite agree with observations: it would predict a too rapid proton decay. Extending SU(5) to SO(10) is very straightforward. The $10 \oplus \bar{5}$ combine with a SU(5) singlet into a 16. The singlet was the right-handed neutrino, which is now also needed to give neutrinos a mass. The 16 is also a fermionic representation. The beauty of the resulting scheme is that quarks, antiquarks, leptons and antileptons all end up in one multiplet, which means that 3-quark systems can decay into leptons, so the proton would become unstable. What I find particularly appealing about this construction is that one ends up with a typically fermionic representation of SO(10), as if nature decided that if something is a fermion in space-time, it better also be a fermion in internal space; thus, with 3 generations, we apparently have three fermions, each forming 32 dimensional (chiral) spinors in a $10 + 4$ dimensional "space-time".

Salam's picture of the natural world was that it should be described by physical laws that are aesthetically pleasing. Beauty was a very important criterion for him. He was enchanted by supersymmetry, supergravity, superspace and superstring theory. These theories had to be true just because they are beautiful. To my taste, he got himself carried away sometimes a bit too easily. When a theory turns out to give unexpected new insights, receiving support from different corners of experimental physics and/or theoretical arguments, it usually also allows us a new view, showing us a new horizon that had been hidden up to that moment. Such new horizons are always beautiful, so indeed, correct theories are beautiful theories, in general. But turning this around very often does not work. It would have been perceived as "beautiful" when matter was made of just four elements, "water", "air", "earth" and "fire". This was the beauty that was searched for by the primitive scientists of the Middle Ages; yet the real truth would turn out to be more beautiful in its own, more complicated ways. This is a lesson that one should always keep in mind. There are beautiful worlds all around us, but one might have to wade through a swamp to reach them.

3. Superstrings and Quantum Gravity

Injecting quantum mechanics into Einstein's theory of General Relativity turned out to be a vastly more stubborn problem than most of us had anticipated. We thought that this was a problem very similar to the riddles we have been facing in the past. Injecting quantum mechanics into Special Relativity had also been hard, and it too at first looked like an impossible assignment. Various approaches using as much experimental input as we could put our hands on, in combination with pure logical reasoning, were tried; and we vindicated: the "Standard Model" was the most precise and complete answer that was uncovered. So should we not do the same thing again, sharpen our theoretical and experimental techniques, produce more precise formalisms, and yes, quantum gravity will be there. Will it? History also shows us that procedures that have proven very successful in the past, do not always guarantee success in the future. To crack this problem, we might first have to step back. And if that does not help, perhaps step back further. By itself, this is a risky thing to say. I receive letters every day from crackpots who give me the same advice: forget all that mathematical gibberish, read the Bible, go meditate, or something of that sort. This is clearly not what I have in mind.

What I do have in mind is that we have to improve our mathematics, but we might have to start at places where we thought everything was secure. My own hobbyhorse is quantum mechanics. It is taken for granted by almost all theoreticians that, in order to formulate some quantized version of gravity, one has to set up a description of the basis elements of Hilbert Space. The first thing we teach to our students is that all linear transformations in this Hilbert space are allowed, and all physical transformations of interest can be reduced to such linear transformations in Hilbert space. There is no particular basis to be preferred in favor of any other.

One might, however, suspect that quantum mechanics is not a completely immune corner stone of all fundamental theories. It might be possible to *explain* why we experience quantum mechanics in the world of the tiny things. Such an explanation might reveal that there is something underneath quantum mechanics. The notion we call Hilbert space is then just degraded into a powerful mathematical machine to handle the stochastic nature of the solutions to some highly complex equations. Could we not try to identify candidates for such underlying dynamical theories? I claim that we can indeed try to do this, and that this may well result is a picture where one particular basis of Hilbert space actually describes ontological reality, whereas others do not. This basis cannot be identified in terms of the particles and fields we know today, because probably the complete set of all particle types, including the ones relevant at the Planck scale only, and the ones describing black holes, have to be included. This will be a gargantuan task, but we can first try to find simplified models where exactly this picture applies. One finds that this idea is not crazy, but we do have to abandon some of our standard procedures. Since we cannot abandon all of our standard procedures at once, this is a difficult path to follow, but it is worth thinking about it.

Compared to what I just said, superstring theory is just a more "conventional" scheme. It is not my intention to criticize superstring theory, but one cannot help noting that something essential appears to be lacking: a precise description of a valid *interpretation* of this theory. Looking at the face of things, one is inclined towards the following interpretation.

What string theory adds to the construct of quantum field theories is not only a notion of one-dimensional — linelike — structures, but also higher dimensional objects, (mem-)branes. There is a whole tower of mathematical features that can be observed here, including general relativity in target space — i.e. gravity in spacetime. It is suggestive that this indeed is the structure that also plays a key role in General Relativity. At the same time, however, there are numerous shortcomings if one wishes to elevate this construct to the level of a fundamental ToE.[a] The structure is fundamentally quantum mechanical, which means that it will never make definite predictions, but only yield statistical expectations for its fundamental variables. What is conspicuously missing is boundary conditions (in space as well as in time), and variables that evolve deterministically. If one believes that Nature (or "God") is perfect, then this "partial" theory is suspect.

The recent "landscape" theories have had a major impact on our "Grand View of Physics". The way the situation is presently formulated appears to lead to a very disappointing state of affairs, since this scheme is not much more than a gesture of surrender: we will never be able to derive the most conspicuous features of the Standard Model. Yet the landscape idea is also difficult to refute. This really could be the conclusion that our quest for understanding will be leading to. The power of superstring theory, together with the impressive mathematical edifice that has been established by its investigators, should not be underestimated. But as long as, in my view, this theory remains incomplete, requiring more solid foundations than what we have at present, there may still be hope.

[a] "Theory of Everything".

ABDUS SALAM AT IMPERIAL COLLEGE

TOM KIBBLE

Blackett Laboratory, Imperial College London, SW7 2AZ, UK

It is a privilege, as well as a great pleasure, for me to talk about Abdus Salam at Imperial College. He is someone to whom I owe a great personal debt. I have always felt that I was very fortunate to have found myself joining his group in 1959, less than three years after he first set it up. It was a very exciting place to be, and a very exciting time in theoretical physics.

1. Salam's Arrival

It was Patrick Blackett, acting I am told on the advice of Hans Bethe, who first invited Salam to apply for a post at Imperial College. He joined the College as Professor of Applied Mathematics, in the Mathematics Department, replacing Hyman Levy. In the same year I joined, 1959, Salam was elected as the youngest Fellow of the Royal Society, displacing Walter Hayman of the same Department. That was where I spent my first year here, as a postdoctoral fellow.

The Maths Department then occupied the old Huxley Building on Exhibition Road, now the Henry Cole wing of the Victoria & Albert Museum. The Huxley Building had been the home of the Royal College of Science before it became one of Imperial's constituent colleges in 1907. It was there that T. H. Huxley worked and H. G. Wells studied. It was a cavernous building with huge rooms, quite unsuited to its modern purpose, into which mezzannine floors and extra partitions had been inserted. I shared with an Austrian-American visitor a rather odd-shaped room that had just half a window, where we could often hear every word of the lecture in the neighbouring lecture room. I'm not sure if they could hear us!

When I arrived, as a NATO Fellow, there were three members of staff: Salam, his right-hand man, Paul Matthews, then a reader, and a young lecturer, John C. Taylor, later a professor in Cambridge. There were a dozen or so PhD students, including Yuval Ne'eman, whose obituary of Salam we heard earlier. They all shared one large room, which was also where we all met for tea or coffee. There was a constant stream of visitors. During that first year they included Murray Gell-Mann, Art Rosenfeld, Gordon Feldman, who became a long-term collaborator of Salam and Matthews, my room-mate, Hans Ekstein, and, remarkably, two visitors from behind the Iron Curtain, Andrzej Trautman from Warsaw and Aston Komar from Moscow. Among later visitors were Steven Weinberg, Stanley Mandelstam, Ken Johnson, Lowell Brown and many others.

The next year, 1960, we all moved into Physics, with Salam becoming Professor of Theoretical Physics — though the move was very much against the wishes of Harry Jones, then Head of Mathematics. But Blackett had at least two strong inducements up his sleeve. One of course was moving into the brand new Physics Building, now the Blackett Laboratory, where we had much better facilities. Another was that the Group was then right next to the high-energy experimental group, led by Clifford Butler.

Fig. 1.1. Professors in the Physics Department, c. 1964. Front row: Abdus Salam, Clifford Butler, Patrick Blackett, David Wright, Harry Elliot. Back row: Paul Matthews, Maurice Blackman, Jim McGee, John Mason. (Blackett Laboratory Photographic Section)

Blackett had made huge changes to the Department since he arrived a few years earlier. When he came, in 1954, there were just three professors. Ten years later, as Fig. 1.1 shows, there were nine, including Paul Matthews as well as Salam; the Theoretical Physics Group was the first to have more than one professor.

The Group had also grown substantially by then; in addition to Salam and Matthews there were three lecturers, including myself, and a large number of

Fig. 1.2. Theoretical Physics Group, c. 1964. Front row: Munir Rashid, John Charap, Tom Kibble, Abdus Salam, Paul Matthews, Mavis Avis, Ray Streater, Arif-uz-Zaman, Ron King. Back row: Ansarrudin Sayed, Yahya Khan, Shaun Dunne, Jimmy Boyce, Ghulam Murtaza, Ian Bond, unknown, Ray Rivers, Ian Yamanuchi, Ian Poston, Sarwar Razmi, John Strathdee, Ian Ketley, Kamaluddin Ahmed, Dick Roberts. (Blackett Laboratory Photographic Section)

students and postdocs, and of course our hard-worked secretary, Mavis Avis. Figure 1.2 shows the Group in about 1964. With the help of colleagues, I have been able to identify almost everyone (I hope accurately), though there remains one (possibly Russian) whose name is unknown.

1.1. *Excitement of QED*

Why was it such an exciting time and place? I hope those of you who know this story will forgive me if I spend a few moments setting the scene.

The end of the Second World War had led to a huge flowering of fundamental physics, as so many of those physicists who had been involved in one way or another in the war effort returned to academia. One of the greatest advances was the development in the late forties of *quantum electrodynamics*, or *QED*, the quantum theory describing the interactions of electrons and photons, the first theory that successfully reconciled quantum mechanics with Einstein's special theory of relativity. The theory had been around in embryonic form since the mid-thirties, but suffered from an intractable problem of infinities. There is a natural small parameter in the theory, the *fine structure constant*, α, a dimensionless number formed

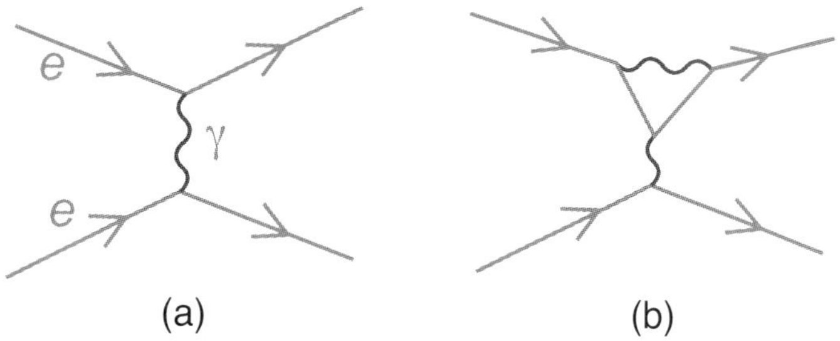

Fig. 1.3. Feynman diagrams for electron-electron scattering

from the charge on the electron e, Planck's constant h and the speed of light c. Essentially, $\alpha = e^2/hc$; using the rationalized (SI) units for charge, we usually write it as

$$\alpha = \frac{e^2}{4\pi\epsilon_0\hbar c} \approx \frac{1}{137},$$

(where $\hbar = h/2\pi$ and ϵ_0 is the permittivity of the vacuum).

Calculations are done in *perturbation theory*; that is to say the results are in the form of a power series in α. The various terms in this series are represented by so-called *Feynman diagrams*. For example the lowest-order result for the scattering of one electron by another is represented by a diagram in which a single photon is exchanged between the two, as in Fig. 1.3(a). Higher-order correction terms are represented by diagrams containing additional loops, such as Fig. 1.3(b). The problem was that while the lowest-order calculations gave perfectly sensible answers, trying to calculate the "small" corrections of higher order in α always yielded infinite answers.

This problem was brilliantly solved independently by three physicists, Richard Feynman, Julian Schwinger and Sin-Itiro Tomonaga, by the device of "renormalization". Among other things the theory predicted infinite corrections to the mass and charge of the electron. What they showed was that the infinities could all be collected into corrections to these "bare" quantities; in calculating the relations between actual physically measurable quantities such as reaction rates and the real measured mass and charge the infinities all cancelled out. So for the first time they were able to compute higher-order correction terms. And the results were spectacular.

At the same time, experimenters had been making measurements of unprecedented accuracy. The new calculations in quantum electrodynamics agreed astonishingly well — in particular, with the *Lamb shift*, the measurement by Willis Lamb of the tiny splitting between two energy levels in hydrogen, the $2s_{1/2}$ and $2p_{1/2}$ levels. Very good agreement was also found with a small correction to the magnetic moment of the electron. It was clear that whatever one might think about the validity of cancelling infinities, the theory really did work. Indeed, this success has

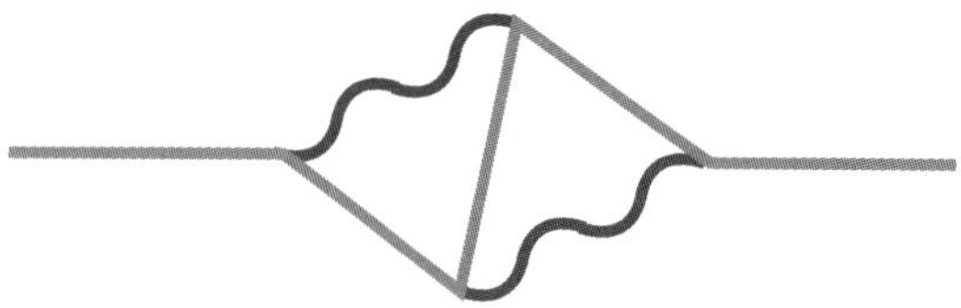

Fig. 1.4. An overlapping divergence.

continued: quantum electrodynamics is now the most accurately tested theory in physics.

Freeman Dyson formalized the procedure, demonstrating that all three approaches are essentially equivalent, and proved that renormalization works to all orders in α. Salam's first contribution that grabbed the attention of the physics community was to fill in a gap in Dyson's proof, concerning the treatment of so-called "overlapping divergences" (see Fig. 1.4).

1.2. *The next challenge*

After the triumph of quantum electrodynamics, the big challenge was to find similarly successful theories of the other kinds of interactions of elementary particles. We generally distinguish four classes of interactions, according to their strength and range. The strength is measured by dimensionless coupling-strength parameters similar to the fine structure constant. There are two long-range forces with which we are familiar in macroscopic physics, both of which obey the inverse square law — the electromagnetic and gravitational — but on the scale of particles they are of hugely different strengths, as is clear from the fact that chemical forces of electromagnetic origin on our feet are enough to counterbalance the gravitational attraction of the entire Earth and stop us falling through the floor. For the gravitational force the analogue of the fine structure constant is $G_{\mathrm{N}}m^2/\hbar c$, where G_{N} is Newton's constant and m is the particle mass. This is tiny, around 10^{-40}.

There are also two classes of short-range nuclear forces, whose effects are negligible beyond a few nuclear radii: the strong nuclear force that binds protons and neutrons in atomic nuclei and the weak nuclear interaction responsible for radioactive beta decay, and indeed for powering the Sun. The coupling strengths for these are typically of order 1 and 10^{-10} respectively.

Initially, most of the emphasis was on the strong forces, and indeed one of Salam's important early contributions (with Matthews) was to show how to generalize the successful renormalization procedure from QED to models involving the exchange of spinless particles. Hideki Yukawa had shown before the war that the short range of strong forces could be understood if the mediating particles, unlike the photon, had a non-zero rest-mass; in fact

$$\text{range} = \frac{\hbar c}{\text{mass}}.$$

The obvious candidates seemed to be the recently discovered π-mesons or *pions*, which are spinless and have masses of just the right magnitude. However there was a serious obstacle to constructing a satisfactory theory of strong interactions: the large coupling strength. In QED we have a nice small parameter. Successive corrections to any physical prediction are reduced in magnitude by powers of α. But for the strong interactions that is not true: higher-order correction terms would be as large as the lowest-order term. So the calculational method breaks down. Many people came to the conclusion that quantum field theory could not be used to describe the strong interactions, and looked for alternatives, in particular so-called S-matrix theory.

There was another growing challenge for theoretical physicists — to try to make some sense of the zoo of apparently fundamental particles turning up in cosmic-ray and accelerator experiments. We started with just the electron, proton, neutron and photon, but by 1960 there were dozens — not only the pions and *muons* or heavy electrons, but also a whole slew of "strange" particles: the K-mesons, the Λ, Σ and Ξ baryons, and then many others.

1.3. *The role of the Imperial College Group*

So what were we doing at Imperial College? Many different things, of course, but there were I would say two main themes.

The first was *symmetries*. Surely, these dozens of new particles could not *all* be elementary? As a first step in trying to understand this proliferation, it was natural to ask: could we find symmetries between them? Could we show that they could be grouped into multiplets of particles with similar properties, thus reducing the number of independent entities?

The second — and in the end Salam's true passion — was *unification*. Could we find descriptions of the other interactions as successful as QED? Or better still, could we find a *unified* theory of the different interactions? — especially a *gauge* theory, about which I will have a lot more to say later.

Let me turn first to symmetries. The first symmetry of this kind, now called "*isospin*", was based on an obervation by Werner Heisenberg in 1932. He noted that the proton (p) and neutron (n) are in many ways very similar: their masses are nearly the same, and so are the strong nuclear forces between them. Heisenberg's idea was that the proton and neutron could be regarded as two states of a single entity called the *nucleon* (N). We could write $p = N^+$ and $n = N^0$, where the superscripts indicate the electric charge. Then in 1936, Breit, Condon and Present drew an analogy between these two *charge states* and the two *spin states* of an electron (e): $e_\uparrow$ and $e_\downarrow$, which are distinguished by the value of the z-component of the spin angular momentum: $S_z = +\frac{1}{2}\hbar$ and $S_z = -\frac{1}{2}\hbar$ respectively.

Quantum mechanics is invariant under rotations of the spin direction; but unlike the situation in classical physics, rotations do not generate distinct states, but only linear combinations of the two. We can write, for example

$$e_{\nearrow} = ae_{\uparrow} + be_{\downarrow},$$

where a and b are calculable numbers. The bold hypothesis of Breit and his colleagues was that the strong interactions should similarly be invariant under "isospin rotations", rotations not in ordinary space but in a conceptual "isospin space", under which p and n are carried into linear combinations, e.g.,

$$p \to ap + bn, \qquad n \to -bp + an.$$

The entire group of such transformations is called SU(2), the *special unitary group* of dimension two.

They showed that this hypothesis leads to predictions about nuclear energy levels that are well fulfilled: the energy levels of nuclear isobars (nuclei with the same total number of protons and neutrons) can be grouped into "isospin multiplets" with approximately equal energies, although this is not an *exact* symmetry — it is broken by the fact that the proton has an electric charge, while the neutron does not. Hence one expects small deviations from energy equality.

Our modern view of isospin symmetry is that it reflects the underlying structure of the particles involved. All the *hadrons* (strongly interacting particles) are composed, we believe, of *quarks*, that come in six different "*flavours*" $(u, d, c, s, t, b$, standing for *up*, *down*, *charmed*, *strange*, *top* and *bottom*), together with their antiquarks. All the long-lived hadrons (including the proton and neutron, and the pions) are made of the two lightest quarks (u and d) together with their antiquarks ($\bar{u}$ and $\bar{d}$). For example, the nucleons and pions have the composition:

$$\begin{array}{ll} p = (uud), & \pi^+ = (u\bar{d}), \\ n = (udd), & \pi^0 = (u\bar{u} - d\bar{d}), \\ & \pi^- = (d\bar{u}). \end{array}$$

Fundamentally, the SU(2) transformations act on the two-dimensional space spanned by these two quarks, u and d.

It was natural to try to generalize the concept of isospin to classify the whole zoo of new particles. The first generalization in fact was made in 1938 by Nicholas Kemmer, who was later Salam's research supervisor in Cambridge. He showed that the three pions could consistently be regarded as an isospin triplet, $\boldsymbol{\pi} = (\pi^+, \pi^0, \pi^-)$, and that one could write down an isospin-invariant form for the strong interaction involving nucleons and pions. For the experts, the form of the interaction is

$$g\bar{N}\boldsymbol{\tau} \cdot \boldsymbol{\pi} N.$$

In 1956 Sakata suggested that all the hadrons might be regarded as bound states of three particles, (p, n, Λ), and then in 1959, Ikeda, Ohgawa and Ohnuki, proposed an extension of isospin, an even more approximate SU(3) symmetry acting on this triplet. This was actually used by Salam, together with John Ward, in 1961 as the basis for an SU(3) gauge theory of strong interactions, though it was soon superseded.

Salam and many of his students worked on a variety of possible extensions of isospin. The one that has stood the test of time is the "*eightfold way*" found by Salam's student, Yuval Ne'eman, and independently by Murray Gell-Mann. It is also based on an SU(3) symmetry, but the fundamental triplet is not (p, n, Λ) but rather the three lightest quarks, (u, d, s). The nucleons and the Λ do not belong to a triplet, but instead to an octet including other strange baryons, the Σ and Ξ. For example, $\Sigma^+ = (uus)$, while $\Xi^0 = (uss)$.

1.4. *Gauge theories*

The origins of the "*gauge principle*" go back to the work of Hermann Weyl in 1929. It is well known that physical results in quantum mechanics are unchanged if the wave function $\psi(x)$ of an electron, say, is multiplied by any phase factor:

$$\psi(x) \to \psi(x)e^{i\alpha}.$$

Weyl asked the question: can this be extended to invariance under a *local* change of phase:

$$\psi(x) \to \psi(x)e^{i\alpha(x)},$$

where the exponent is a function of position? The answer is yes, *provided* that we also include an interaction with the electromagnetic field, which has to transform in a special way under these "gauge transformations". The precise form of the interaction between the electron and this "gauge field" is fixed by the requirement of gauge invariance.

The next major step was taken in 1954 by Yang and Mills who suggested a gauge theory of strong interactions, based on the idea of applying the gauge principle not to phase transformations but to isospin rotations. They showed that this would require a triplet of gauge-field particles, forming an isospin triplet with ordinary spin equal to one (in units of $\hbar$). The same model was proposed independently by Ronald Shaw, a student of Salam's in Cambridge. While this bold idea turned out not to be the correct theory of strong interactions, it provided a blueprint for all the later models. Salam was convinced from an early stage that the fundamental theory of all the particle interactions would turn out to be a gauge theory.

But there was a major problem: the theory seemed to require that the new gauge particles, like the photon, should be massless, in which case they would mediate an infinite-range interaction, quite different from the short-ranged nuclear forces.

Another problem, mentioned earlier, was that perturbation theory is problematic for the strong interactions, because the dimensionless parameter that describes their strength is of order one, rather than 1/137. So even though Yang and Mills, and Shaw, could write down a plausible gauge field theory of strong interactions, no one could calculate its predictions, so there was no way of testing it. For this reason,

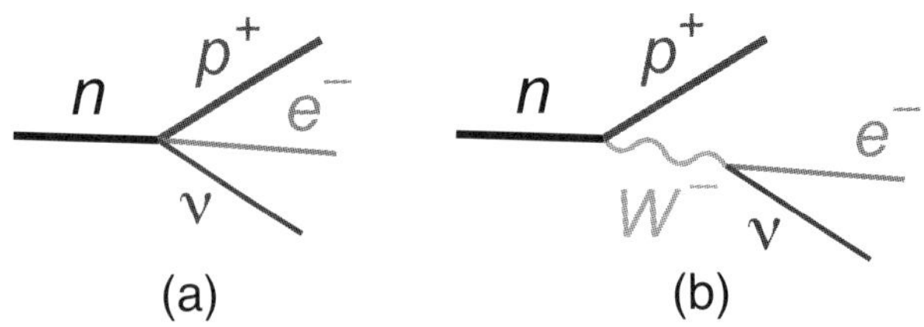

Fig. 1.5. (a) Fermi's view of neutron decay; (b) the intermediate vector boson description.

the weak interactions began to seem more promising, especially after the realization that they too could be well described by a theory involving vector (spin-one) bosons. This *intermediate vector boson theory* was proposed independently in 1958 by Robert Marshak and George Sudarshan and by Richard Feynman and Murray Gell-Mann.

The earliest description of weak-interaction process, due to Enrico Fermi, envisaged a process such as neutron decay (into a proton, electron and neutrino) as occurring via a direct four-particle interaction, as pictured in Fig. 1.5(a). In the intermediate vector boson (IVB) model, it was mediated instead by a vector boson W^-, as in Fig. 1.5(b), giving a picture much more similar to a QED process. The proposers of the model pointed out that the short range of the weak force required that the mass of the W^- be of order 100 GeV/c^2, and that this large mass would also give the correct overall coupling strength if the fundamental three-particle couplings were of the same order as the fine structure constant.

This model immediately suggested to Salam and others that the weak interactions should be described by a gauge theory, and even raised the very interesting question of whether it might be possible to formulate a *unified* theory of electromagnetic and weak interactions. Indeed a rather crude model of this kind, involving a triplet of mediating fields, W^+, γ, W^-, was proposed by Salam and Ward in 1958.

There were, however, two major stumbling blocks. Firstly, the masses of the $W^\pm$ had to be very large, while the photon is massless. How could they then belong to the same multiplet?

Then there was the problem of parity violation. Some odd features of weak-interaction processes led Lee and Yang to suggest in 1957 that they violate parity conservation, meaning that they lack mirror symmetry: a weak-interaction process will in general look different if viewed in a mirror. This was verified in the same year in some beautiful experiments by Wu and collaborators at Columbia University. They studied the beta decay of cobalt-60: $^{60}\text{Co} \to {}^{60}\text{Ni} + e^- + \bar{\nu}$. Cobalt-60 was chosen because it has a large nuclear spin, and thus it is fairly easy to use a magnetic field to ensure that all the cobalt atoms are aligned with their spins in the same direction. The experiment looked at the correlation between this direction and the direction of the emitted electron. They found there *was* such a correlation: the

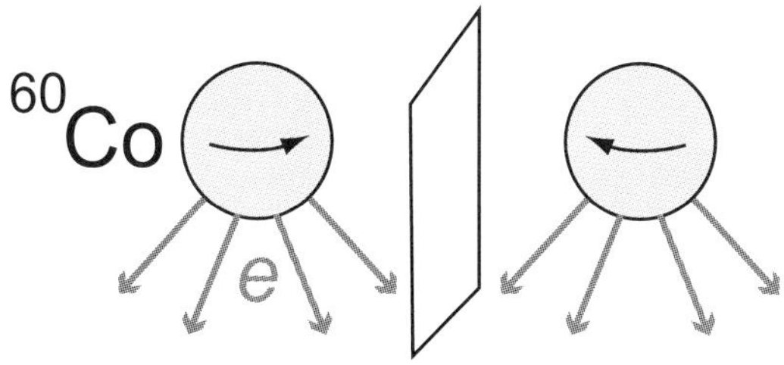

Fig. 1.6. The decay of cobalt-60, viewed directly or in a mirror.

electrons were preferentially emitted in the direction opposite to the spin. But as is clear from Fig. 1.6, the situation would appear reversed if viewed in a mirror.

This immediately leads to a severe problem for a unified theory: how is it possible that the mediating gauge particles can belong to the same multiplet, when the $W^{\pm}$ induce parity-violating interactions while the electromagnetic interactions, mediated by the photon, are parity-conserving, with no left-right asymmetry?

2. Symmetry Breaking

The problem here is that we were searching for a unified theory, in which there is some kind of symmetry between the photon (γ) and the weak intermediate vector bosons $W^{\pm}$. But in two major respects, mass and parity, they do not *appear* symmetric. So if there is such a symmetry it must somehow be *broken*. A very interesting idea that began to attract a lot of attention was that this might be an example of a *spontaneously broken* symmetry. This occurs when there is an underlying symmetry, but the symmetric state is for some reason unstable. A simple and oft-quoted example is a pencil standing on its tip. This set-up has perfect rotational symmetry, but the symmetry will not survive: the pencil is bound to fall in some random direction.

So the question was: could there be an underlying symmetry between the γ and $W^{\pm}$ that broke spontaneously to generate the observed differences in mass and parity conservation?

But this idea ran into another apparent brick wall, the *Goldstone theorem*, initially suggested by Jeffrey Goldstone and proved in 1962 by him, together with Salam and Weinberg. Briefly, it states that the spontaneous breaking of a continuous symmetry necessarily implies the existence of scalar (zero-spin) *massless* particles, corresponding physically to waves in the direction of the symmetry breaking. No such particles were known; nor, indeed, have any been found since then. So it seemed as though instead of solving the problem of massless *vector* bosons, this mechanism would make things worse by adding massless *scalar* bosons as well.

The solution to this problem appeared in 1964 in the form of the *Higgs mechanism*. It became clear that there was a possible escape from the proof of the Goldstone theorem: it does *not* apply to gauge symmetries.

The initial spark that led to this discovery was due to Schwinger. It was almost universally believed that the photon has zero mass *because* it is a gauge boson. But Schwinger showed in 1962 that in fact a gauge symmetry *need not* imply that the corresponding gauge vector boson is massless. Explicit examples of this were provided by Philip Anderson in 1963. He pointed out that the photon itself acquires a mass in a plasma, given by

$$m_{\text{pl}} = \frac{\hbar\omega_{\text{pl}}}{c^2} = \frac{\hbar}{c^2}\sqrt{\frac{e^2 n_e}{4\pi\epsilon_0 m_e}},$$

where ω_{pl} is the plasma frequency, n_e is the number density of electrons in the plasma, and m_e the electron mass. He also noted that in a superconductor, where there is spontaneous breaking of the electromagnetic gauge symmetry, the massless gauge bosons and massless Goldstone bosons in a sense "cancel out", combining to form massive gauge bosons.

There was still doubt as to whether such a mechanism could work in a fully relativistic theory, but relativistic versions of the mechanism were described in 1964 by Peter Higgs, by François Englert & Robert Brout, and independently, a couple of months later by Gerald Guralnik, Richard Hagen (both visitors to Imperial College) and myself. So this at least provided an escape from the first of the two major stumbling blocks, the unwanted massless gauge bosons.

There remained however the problem of parity. If there is a triplet of gauge bosons, W, and the photon is identified with its neutral member ($\gamma = W^0$), how could it have parity-conserving interactions when the $W^{\pm}$ do not? The solution was originally proposed by Sheldon Glashow in 1961, though *without* the crucial element of spontaneous symmetry breaking. This was added in 1967 by Weinberg and independently, in 1968, by Salam. The key points are these:

- the photon is *not* identified with W^0;
- there are not three gauge bosons, but *four*, a triplet (W^+, W^0, W^-) and a singlet B^0;
- correspondingly, the symmetry is not merely SU(2), but SU(2)×U(1);
- the photon is a linear combination of the W^0 and B^0, $\gamma = B^0 \cos\theta + W^0 \sin\theta$ — here θ is the "weak mixing angle";
- the orthogonal linear combination, $Z^0 = -B^0 \sin\theta + W^0 \cos\theta$ is another gauge boson, which mediates "neutral-current" weak interactions;
- the pattern of symmetry breaking ensures that the $W^{\pm}$ and the Z^0 acquire large masses, while the photon remains massless.

Salam named this the *electroweak* theory. It was in fact the detection of neutral-current events at CERN in 1973 that provided the crucial experimental confirmation of this model, and eventually persuaded the Nobel Committee to award the 1979 Physics Prize to Glashow, Salam and Weinberg.

3. Further Developments

Of course this was not the end of the story, though it brings me to the end of what I can cover in this talk. So far as the electroweak theory was concerned, there were at least two further key developments. In 1971, Gerard 't Hooft proved that like QED the theory is renormalizable; Salam and Weinberg had conjectured that might be true but had not been able to prove it. This result stimulated widespread interest that had earlier been lacking, and encouraged CERN to start their experimental tests.

Then in 1983, also at CERN, came the actual detection of the $W^{\pm}$ and Z^0 particles.

So far as Salam's own work was concerned, the success of electroweak theory led on to attempts to incorporate the strong interactions and eventually gravity too. With Jogesh Pati he developed in 1973 one of the first "grand unified theories", using a bigger symmetry group to tie together the electroweak and strong interactions. A key prediction of their model was that the proton should be unstable, albeit with a lifetime many times greater than the age of the Universe. Although this particular model has been superseded, it had a great influence on later work.

Salam also took a great interest in the idea of *supersymmetry* — an extended kind of symmetry that could for the first time group together fermions, the constituents of matter, with bosons, the carriers of interactions. This is an essential part of modern superstring theory (or "M-theory"). Salam made a key contribution to this work in 1974, developing with John Strathdee the "superfield formalism" which made many calculations more tractable. And there was much, much more, along similar lines.

But I have omitted altogether all his activities outside of physics. From our perspective, in his group at Imperial College, he seemed to be prodigiously productive in research, always full of new ideas. But at the same time, he was doing many other things, that we were only dimly aware of. From 1961 to 1974 he was Scientific Advisor to the Government of Pakistan, appointed initially by President Ayub Khan. He was the Pakistani delegate to the General Council of the International Atomic Energy Agency (IAEA), and a tireless advocate of the importance of science for developing countries.

Salam deeply regretted having to leave Pakistan to pursue his chosen career, and wanted to help others in a similar position to avoid having to make this choice. He persuaded his fellow delegates at the IAEA to back his scheme for setting up an International Centre for Theoretical Physics (ICTP) — an astonishing achievement considering that it was initially opposed by almost all the major powers. He obtained backing from the Italian government and the city of Trieste, including the provision of a building and a promise of $200,000 per year for four years.

The Centre — now the Abdus Salam International Centre for Theoretical Physics — opened in 1964 in a building in Piazza Oberdan in the centre of the city, a building which, I understand, was once the home of the gestapo in Trieste.

The ICTP moved to a fine new purpose-built building near Miramare on the outskirts of Trieste in 1968, and has been a huge success, providing a lifeline for many scientists from the third world.

From our narrow viewpoint at Imperial College, this success did have a downside. From the mid-sixties onwards, Salam spent increasingly more of his time in Trieste, though he always maintained his links with the College, and spent a few months of the year here. But of course on the credit side, many of us enjoyed our visits to Trieste!

4. Salam's Legacy

Abdus Salam achieved so much, in so many fields, that it is difficult to summarize his legacy. I can only give a personal selection of highlights.

He left us with some beautiful physics; his name will live on for a long, long time in the annals of the subject.

At Imperial College, he developed one of the leading theoretical physics groups in the country, with a very strong international standing. It is still going strong.

In Trieste he set up a pioneering and hugely successful international centre — also still going strong — that has been the model for many others around the world.

But above all, he was a man of great humanity, who left behind him a very large number of grateful friends and colleagues, for whom he was a powerful inspiration.

MEMORIES OF WORKING WITH ABDUS SALAM

CHRISTOPHER ISHAM

Faculty of Natural Sciences, Imperial College London, London SW7 2AZ, UK

1. Introduction

It is a great pleasure to have this opportunity to reminiscence about my experiences of working closely with Abdus Salam. He was a remarkable man, and I remember him with the greatest respect and affection.

The story begins at Imperial College in the mid 1960s. My younger colleagues may be shocked to learn that there was then no QCD, standard model, supersymmetry, loop quantum gravity or, most grievous of all, no superstring theory! Instead, the dominant paradigm was S-matrix theory, whose goal was the construction of particle scattering-amplitudes using postulated complex-analytic properties.

Cambridge was the focus for this activity in the UK, although much work was done at Imperial College too. However, the main thrust in our Group was quantum field theory. Worldwide, the lack of any renormalisable theory of the weak or strong interactions had resulted in a general demise of interest in perturbative quantum field theory. With the exception of quantum electrodynamics, and the axiomatic Wightman programme, quantum field theory was largely viewed as being just a tool with which to construct "phenomenological" models of current algebra. However, both Paul Matthews (then the Head of our Theory Group) and Abdus Salam kept the full QFT-flag flying: somewhat to the disapproval of our colleagues in Cambridge I think!

At that time, Abdus Salam was already the Director of the International Centre for Theoretical Physics. The Centre had been recently rehoused in a brand new building in Miramare, overlooking the Adriatic, a few miles outside the city of Trieste. Salam[a] commuted regularly between Trieste and London, working with

[a]As a sign of our respect for him, John Strathdee and I always addressed Abdus Salam as "Professor Salam". However, when writing about him it is difficult to decide on what is appropriate. To keep saying "Professor Salam" is rather clumsy, and in the modern Western world it would be natural to just call him "Abdus". However, he told me several times how much he disliked this, except when used by special friends and colleagues like Paul Matthews and Tom Kibble. For that reason, I have adopted the old, traditional English convention of using what, if he had had one, would be just his surname, viz "Salam". For me, this is a respectful mode of address.

his long-term collaborator in Trieste, John Strathdee, and with Bob Delbourgo at Imperial College.

I entered the scene in 1966 when I started a PhD under the supervision of Paul Matthews. Salam was viewed with awe by us students, and I spoke to him only a few times. Paul Matthews was a great supervisor and produced a fascinating project that was productive enough for me to get a lectureship in 1969, just as I was finishing my PhD. Paul felt, probably wisely, that the transition from student to, very young, staff would be eased if I went away for a year, and I was duly dispatched to Trieste to collaborate with Salam.

2. The Art of Inductive Proofs

I can truly say that working with Salam was a character-building experience. He had a complex, many-faceted personality, and worked in many areas, and in many ways. This included being a creative scientist; the Director of the Center; and a diplomat on the international stage, representing the poorer nations. To all these activities he brought an unbounded energy and optimism.

However, Abdus Salam had many other interests too, and only slowly did I realise that to each individual he revealed only a fraction of his total personality. This was brought home to me at a meeting at Imperial College that was held in his memory shortly after he died. Tom Kibble and I both spoke, as did a number of other people from outside Imperial College, representing Salam's various interests and activities. I remember with particular clarity one speaker saying that "Salam was always angry", which truly astonished me. True, some times he would be a bit irritable after jetting in that morning from some distant part of the world, but to me he was invariably friendly and kind.

One of the reasons why it was such fun to work with Salam was his great sense of humour, imbued with a strong sense of ironic self-awareness. The secretaries frequently spoke of hearing loud laughter rolling down the almost empty[b] corridors of the Centre. Begin able to laugh at oneself is a great help in life, and this is one of the most valuable lessons that I learnt during my collaboration with Salam.

The rolling laughter that invaded the distant secretarial offices had its roots in Salam's rather scatter-gun approach to research. He was a font of original ideas, but did not always have the best judgement about what was likely to be fruitful, and what was not. Throughout his long, and distinguished, career at the Centre, John Strathdee was Salam's primary filter in this respect.

The methodology of most scientists is to do some research and then present the results in a paper. However, Salam frequently reversed these steps. His way of launching a new research programme was to hastily scrawl out a document consisting of just an Introduction that announced some great new result, and a

[b]As I recall, that winter of 1969 there were only half a dozen or so physicists in the Centre, and about twenty administrative staff.

Conclusion that reiterated the same. He would then give John and I the task of proving this result, and then completing the paper. It was the intrinsic bizarreness of this approach that generated much of the good humour, although it could sometimes put pressure on John and I to prove something that we were certain was false:-)

This occasional feeling of pressure was enhanced by Salam's singularly individual, tripartite classification of theoretical physicists into "youths", "tomtits", and "broken reeds".

The highest achievement was to be a "youth": an ascription that Salam reserved for those for whom, sometimes grudgingly, he had respect. Thus, for example, Murray Gell-Mann and Wigner were "youths", as was Tom Kibble; so you can get the general idea! The exception was Paul Dirac, who was Salam's idol and for whom he had the utmost admiration. Dirac always stood outside the tripartite scheme.

I can only hope that Salam eventually regarded me as a youth, as to be labelled a "tomtit" was not what one would wish. This term was reserved for anyone he thought was not up to scratch, but particularly for those whom he regarded as being, what would these days be called — somewhat impolitely — a "bull-shitter". I can see one or two of Salam's nominated tomtits in the audience this evening, but prudence suggests I should leave it at that!

But the worst thing for John or I would have been to be labelled a "broken reed" as Salam kept this term almost exclusively for former collaborators who had fallen beneath his expectations and cracked under the strain. However, quite a few of these people went on to have good careers in academia, so clearly his standards were rather high!

I had a peculiar problem in this direction because Salam regarded me as the most mathematical of the three of us, and to me therefore would fall the task of proving some of his more "optimistic" speculations. This was not helped by his version of induction. Normal induction reasoning allows you to prove a sequence of propositions $P_1, P_2, \ldots$ by showing that (i) for all $n = 1, 2, \ldots$, if P_n is true then so is P_{n+1}; and (ii) P_1 is true. However, Salam's more "liberating" version is captured in his famous "three laws of induction" which, in increasing order of power, are:

> **Salam's first law of induction:** If P_1 and P_2 are true, then P_n is true for all n.
> **Salam's second law of induction:** If P_1 is true, then P_n is true for all n:
> **Salam's third law of induction:** If the propositions P_n, $n = 1, 2, \ldots$ are physically interesting[c] and would make a good paper, then P_n is true for all n.

My own experience of doing research in theoretical physics has been that Salam's first law works surprisingly often. However, applying the second law requires considerably more insight, and the less said about the third law the better:-)

[c]In its original form, this law referred only to "making a good paper". The cautionary qualification "... are physically interesting and ..." was added by John. For this reason, Salam's third law is sometimes known as "Strathdee's first Lemma".

In a similar spirit to his laws of induction was Salam's general maxim that, in doing research, "There is no such thing as a failure". Thus he greeted each result with triumphant exclaim, even if it was not what he had been originally expecting. I remember once when we were writing a paper about spontaneous symmetry breaking of the space-time conformal group. Somewhat implausibly in retrospect, the conjecture had been that the Goldstone bosons for this process would have a mass. Then, just as the Conclusion was being finished, John and I discovered a mistake in our calculations and found that these bosons were actually massless. We naturally assumed that this was the end of that particular research project, and were rather gloomy. However, Salam felt otherwise, and with a deft editorial hand transformed the paper into a triumphal acclaim of the proof of the masslessness of the Goldstone bosons for a space-time symmetry group! The idea that "there is no such thing as a failure" is something that I teach to my students to this day.

One of the great joys of working with Salam were the humorous remarks and gestures that would accompany his invocation of the third law of induction. I can see him now, sitting back in his chair, twirling his moustache, and saying, with a twinkle in his eye "Ah, Professor[d] ...", a pause, and then "it is clear that $1 = 2$", or whatever else it was that he was trying to bamboozle me about. If, finally exasperated, I would demand that he actually prove that $1 = 2$, or whatever, he would pronounce solemnly, but with eyes still twinkling, "I am a humble man", the implication being that he was not worthy enough to aspire to the heights of actually providing a proof. If even that failed to move me, he would touch his nose with his right hand[e], and point to the sky with his left: a traditional gnostic gesture for invoking the ultimate deity. After that, I could say nothing:-)

Another favourite trick of Salam was to give John and I some incredibly difficult task that would have taken us many months to complete, go away for a couple of weeks, return, come into the office we shared, rub his hands in joyful anticipation, and say "All done?" Of course, this was mainly a joke, but not entirely because ...

3. Salam the Hard Worker

Abdus Salam was an incredibly hard worker! He would rise early each day and start work immediately. When I was back in London, there were more than a few occasions when I staggered out of my bed at 5am (literally) on a Sunday morning to anxiously answer the phone, fearing that some disaster had befallen someone close to me, only to hear a familiar voice saying "Ah, Professor, ...". He would then bombard me with lengthy equations that I was supposed to manipulate in my

[d]He frequently addressed me thus when he was in a good mood, and was trying to persuade me of the truth of some dubious-sounding conjecture he had just made. Of course, in those days I was really just plain "Dr" and, within the British system, a long way from earning the title "Professor"!

[e]I should confess that I have inherited these gestures from Salam, and for decades now my students and collaborators have suffered under them at my hand!

head: difficult at any time, and impossible at 5am on a Sunday! He would use this technique with particular gusto when he was trying to convince me that 1 = 2:-)

But my best example of Salam's working habits occurred in the summer of 1971 when I had returned to Trieste for a few months and was staying in, what was known as, "Fronsdal's villa"[f] which was right next to the Centre, and therefore to Salam's own villa. That summer, John Strathdee was on a visit to New Zealand (the home country of his wife, Barbara) and I was left to work with Salam alone. My daily schedule was firmly fixed for me. I would rise at 7, have a quick breakfast, and then meet Salam in his office at 7:30. We would work continuously until 9:30 in the evening; seven days a week! This schedule did include a short lunch "break" but it wasn't really a break as we always discussed work throughout it (there was one exception to which I will return later).

This schedule was hard work, but also posed one particular domestic problem for me. Summer in Italy is hot, and shirts and socks need to be washed rather frequently. If I have one complaint about working with Salam it is that he never seemed to grasp the fact that I really did need to take off the odd half-hour to perform this task. I presume someone else always did this particular job for him and, unlike me, he did not need to hastily clean his underclothes in a wash-basin late at night:-)

I was exhilarated by this experience of working continuously for 14 hours a day, but it was also exhausting, and it was probably a good thing that after around a fortnight Salam would be off somewhere, and I could recover!

4. Salam and the Art of Management

It was during this same trip that Salam decided that I needed some instruction in the art of management. He was probably right since, in those days, I was still young and innocent. Of course, nowadays, we are all "line managers", and the ability to be such is a requirement for any academic post. But I joined the academic world when a university was still a university, not a commercial business.

Something that struck me forcibly about Salam was his understanding of the Western academic psyche, particularly our characteristic habit of thinking that we are rather better, and more important, than we actually are. I remember with great clarity one day when we were walking down to lunch, and a moderately well-known Italian experimental physicist was just ahead of us. Salam told me that he wanted this man to help in some way vis-a-vis the funding of the Centre, and that he would show me how one went about securing such a thing. He walked up to the scientist, put his arm around his shoulders, and said "Professor X, you are the greatest living experimental physicist since Rutherford!". Professor X puffed up, looked suitably gratified at this well-earned recognition, beamed, and responded "Si: certo, certo ..."; and Salam got the assistance he required:-)

[f] Not unreasonably since the villa was owned by Christian Fronsdal.

5. Sunday Lunches

But if you ask me "how do you *really* remember Abdus Salam?", my answer is not as an insightful manager, or even as a very creative scientist, but rather as an unusual, and very warm human being.

For me, a special example of this was the one exception to our hectic working schedule. Each Sunday, at around 1pm, we stopped work for precisely two hours, and the two of us would stroll across the road to a restaurant in the park that faced the Centre. The restaurant overlooked the Adriatic, and even now, over 35 years later, I can recall the scene vividly: the Italian summer sun; the sparkling water of the Adriatic; the vines growing over the outside of the restaurant; the smell of grilled fish; and the happy chatter of Italian families eating together.

There were two especially nice things about these lunches. The first was that Salam always paid, which I much appreciated. Of course, Salam never drank any wine although he invariably offered me some. Out of courtesy to him I always declined, although it must be said that grilled sardines do go down better with the odd glass of vino bianco:-)

However, what I really appreciated about these special occasions is that we never talked about theoretical physics. Instead, we discussed a wide range of topics including politics, especially of the Third World; the problems faced by the Centre; and, somewhat surprisingly, theology.

Salam seemed particulary interested in the last subject and would regard me as a spokesman for the entire Christian world. This was a tad unreasonable as I did not know any theology in those days; neither did I regard myself as a practising Christian. However, there was one occasion when I did manage to satisfy my interlocutor. I recall that, just as I was settling down to a nice tiramisu, he suddenly banged his hand on the table and said, quite forcibly, "What does the "Trinity" mean? It makes no sense to me!" Of course, the Islamic religion does not recognise that concept, and, to be honest, most Christians today would be hard put to give an intelligible explanation of how the three Persons — Father, Son, and Holy Ghost — are supposed to relate to each other. However, with a burst of inspiration, I replied that it was like the angular-momentum commutation relations, $[J_x, J_y] = iJ_z$ plus cyclic permutations. Salam was very amused by this, and I passed that particular test!

But what Salam seemed to enjoy most of all was talking about the poetry of the Middle East. He had an extraordinary knowledge of this, and delighted in reciting long poems that he knew by heart. Since these were in the original language, I could not understand a word. However, he would translate them to me with great gusto: probably because they seemed invariably to be about battles in the past where Muslim armies overwhelmed the Christians:-)

6. Epilogue

I knew Abdus Salam from 1969 until his death in 1996, and I always held him in the greatest respect. I moved away from Imperial College in 1973, and our close collaboration came to an end. Although I returned in 1976 we never again worked together, although he always regarded me as his personal encyclopedia of all things mathematical; and, of course, the spokesman for Christian theology.

I developed an intense loyalty and affection for Salam, and one of my most treasured possessions is a handwritten letter from the summer of 1972. I was in Trieste again, but ill health forced me to return home. In his letter, Salam addresses me with great kindness and, I would venture to say, real affection — something that I have valued enormously to this day.

Salam's terminal illness was a great trial for him, and his memory slowly faded. I recall with great sadness when, shortly before he died, he came to a reception at Imperial College, and when I went to speak to him, he did not recognise me at all.

But that is not what I remember. What I remember is the twinkling eyes and twirling moustache; the Italian sun sparkling on the waters of the Adriatic; the pasta, tiramisu and grilled fish; the chatter of Italian families; and Professor Abdus Salam introducing his young collaborator to the problems of the world. All this was a truly formative experience for one so early in his academic career. It was something I will never forget.

IN MEMORIAM: ABDUS SALAM (1926–1996)*

K. S. STELLE
Theoretical Physics Group, Imperial College London,
London SW7 2AZ, UK

The 20th century has been one in which the science of physics has made tremendous progress, starting with the two great conceptual revolutions of Relativity (1905/1915) and Quantum Mechanics (1925). At the same time, however, entirely new domains of physical phenomena have been discovered and explored. These new domains have themselves also provided the tools for exploration, namely radioactive emissions, cosmic rays, and particle accelerators. Molecules, atoms, nuclei, particles, quarks — layer after layer of matter has been revealed and understood. Great new questions, however, have thus also emerged: What are these various constituents, at each level, why just these — and what are the forces acting between them, how are they induced, how do they depend on the constituents' mutual distances and velocities? The answer is given, at each level, in terms of such relevant parameters; yet it is, in fact, a collective "effective" force, really the integrated sum of forces at the level of the next immediate layer and its corresponding constituents. Thus, the most important and the most interesting description would be a reduction down to the "fundamental" layer, if there be one.

It is in this picture that Abdus Salam made his impact, using his physical intuition, his creative imagination and his deep belief that the universe is ruled by laws of harmony and symmetry. He was among the first to emphasize symmetry. His first important contribution was more of a technical nature. Yukawa had suggested in 1935 that the force binding protons and neutrons into nuclei was due to the exchange of a new particle, the pion — which was discovered in 1947. Around that time, the action of the electromagnetic force, binding electrons to the nucleus in the atom, was first understood at the quantum level — even though it had been fully understood at the macroscopic level since the days of Faraday and Maxwell, in the

*Obituary for Professor Abdus Salam written in Arabic by Professor Yuval Ne'eman, Tel-Aviv University, and presented by Professor K. S. Stelle. The Arabic text is reprinted on pages 31–34.

1860s. The breakthrough at the quantum level was due to Tomonaga, Schwinger, Feynman and his student Dyson. The latter was young Salam's first instructor and Abdus' first important result, around 1951, was to extend the new method of "renormalization" to the pion force. Though important in principle, this result was not utilized because the method could not be exploited for such a strong force — about 1000 times stronger than the product of the electrical charges of the proton and electron, in Coulomb's law, for their binding in the hydrogen atom, for instance. In this case, the exchanged "mediating" particle is the photon, a quantum of the electromagnetic field. Note that it is massless and has an unlimited range (our telescopes observe photons arriving from the confines of the Universe), whereas the pion is massive and has a limited range (happily, otherwise we would all be glued in one huge ball).

I was Salam's student in 1958–61, and his belief in the role of symmetry greatly encouraged me in my own project. "You are embarking on a highly speculative project, but if you do it, do it in depth, do not be satisfied with the little Group Theory I know and taught you!" was his advice. As a result, I plunged into the mathematics of group theory, the tool with which to study symmetry. The result was the identification of the pattern displayed by 100 different nuclear particles, many of which had just been discovered. By 1962, I had also began to understand the underlying structure — they were all combinations of three basic "bricks", later named "quarks". Salam improved his technical knowledge from my reports on Group Theory.

It is around 1968 that Salam achieved his most important breakthrough. He had become interested in symmetry breakdown, publishing a first important contribution with Goldstone and Weinberg in 1962. In Salam's group, Peter Higgs and Tom Kibble had made progress with the question of the acquisition of mass by photon-like mediating mesons, through a mechanism in which the relevant symmetry is broken "spontaneously". Glashow had originally identified the symmetry relating electromagnetism to the "weak" nuclear interaction, yet another short range force, responsible for various radioactive decays, first identified as a separate force by Fermi in 1935. Since the photon is massless, whereas the conjectured mediators of the weak interaction were required to be very massive (it acts at very short range), this was a case of symmetry breakdown. Salam then "took the bull by its horns" and at an international meeting in Göteborg in the summer of 1968, presented a full theory of the unified weak and electromagnetic interaction (since called "electro-weak"). He used the Glashow symmetry, the Higgs-Kibble mass-acquisition mechanism, added his own physical picture — and presented a complete theory, the first example of unification since Maxwell had produced a unified theory of electricity, magnetism and light, 100 years earlier. The same result was independently suggested by Weinberg. The first important predictions of the new theory were confirmed in 1978 and Glashow, Salam and Weinberg shared the Nobel Prize in Physics in 1979. The massive mediators of the weak interaction were discovered in 1982.

Salam went on producing many new theories and new predictions throughout the 1970s, mainly a further unifying scheme (with Pati) linking the electroweak force to the strong nuclear interaction (including the pion's action), now seen as acting between quarks. He also suggested that this interquark force might have an additional, gravity-like component ("strong gravity"). These remain to be checked experimentally and theoretically in the coming years.

Salam's contribution to the science of physics is only one face of his activity. Aware as he was of the importance of science of development, both economic, social and cultural, he initiated the International Centre for Theoretical Physics at Trieste (1962), under the auspices of UNESCO and the International Atomic Energy Authority (IAEA). This has become the main school at which the countries of the developing world have found the means to train their best scientists and achieve contact with the most advanced programs — without losing them in the process. Another of Salam's initiatives in this field has been the creation of the Third World Academy. He also devoted much of his energy to the development of his native Pakistan, sometimes having to overcome difficulties, as a member of the Ahmadiya.

Salam was one of the great leaders of the physics of particles and fields, in the second half of this century. His legacy to the 21st is a lesson in the fundamental role of symmetry and harmony.

البشرى
المجلد ٥٤
نيسان- حزيران ١٩٩٧
الأعداد٤-٦

بسم الله الرحمن الرحيم

نحمده ونصلي على رسوله الكريم

البشرى

مجلة دينية شهرية

تصدرها الجماعة الاسلامية الاحمدية بالكبابير

مدير البشرى ومحررها	هيئة التحرير
المبشر الاسلامي الأحمدي	عبدالله أسعد عودة ـ الحاج موسى نايف سرور
محمد حميد كوثر	فلاح الدين محمد عودة

Al-Bushra (Good News): Monthly Arabic Religious Organ of the Ahmadiyya Muslim Community (f. 1935) P.O.B 6088,

Alkababir- Haifa- TEL - 972-4- 8385002 - FAX- 972-4-8371612

العنوان: شارع رعنان ٦٠ الكبابير - حيفا - ص.ب. ٦٠٨٨ هاتف ٨٣٨٥٠٠٢-٤-٩٧٢ - فاكس ٨٣٧١٦١٢-٤-٩٧٢

المجلد ٥٤	ذو الحجة- صفر	شهادة- احسان	نيسان- حزيران	١٩٩٧ م
الاعداد ٤-٦	١٤١٧-١٤١٨هـ.ق	١٣٧٦ هـ.ش	ابريل- يونيه	

محتويات العدد

في ذكرى المغفور له البروفسور عبد السلام

وصلتنا من البروفيسور يوفال نيئمان الرسالة التالية: في نعي فقيد الجماعة الغالي البروفيسور عبد السلام الذي وافته المنية مؤخرا. ونظرا لأن هذه الرسالة قد وصلتنا بعد اعداد عدد البشرى الخاص بعبد السلام فاننا ننشرها في هذا العدد شاكرين للبروفسور نيئمان تعازيه هذه وراجين المعذرة لنشرها بشيء من التأخير، واليكم الرسالة:

ذكرى عبد السلام (١٩٢٦-١٩٩٦)

لقد كان القرن العشرين من القرون التي خطت فيها علوم الفيزياء خطوات جباره، بدءا بمفهوم النسبية الثوري (١٩٠٥-١٩١٥) ثم ميكانيكيات الكم (١٩٢٥)، مع ذلك فقد تمت في عالم الفيزياء اكتشافات في مجالات جديدة كلية. وهذه المجالات كونت بحد ذاتها آليات لاكتشاف اشعاع الراديو ثم الأشعة الكونية ومعجلات الجسيمات. وهكذا جاءت هذه الجزئيات والذرات والنوويات والجسيمات طبقات الواحدة بعد الأخرى من طبقات المادة. ولكن ظهرت وبرزت معها تساؤلات كثيرة وكبيرة. فما هي المركبات في جميع هذه المستويات وما هي بالذات؟ وما هي القوى الفاعلة فيما بينهما؟ وكيف تتأثر؟ وكيف تعتمد على المسافات والسرعة فيما بينها ، مع ذلك فان هذه القوى هي في الحقيقة قوة ذات فاعلية اي مجموعة قوى الواحدة تلي الأخرى، لذا فان أهم وأفضل وصف لها هو تقليصها الى الطبقة الأساسية ان صح وجود طبقة كهذه.

في هذا المجال ومن هذا المنظور برز تأثير سلام، الذي بذكائه وبديهته الفيزيائية وخياله الخلاق ثم ايمانه العميق أعلن أن الكون تحكمه قوانين متناسقة ومتجانسة فكان سلام من بين الاوائل الذين أكدوا على فكرة التناسق والتناسب، وقد اتسم اسهامه هذا بطبيعته التقنية.

لقد كنت أحد طلاب عبد السلام في السنوات ١٩٥٨-١٩٦١ فكان ايمانه العميق في نظرية التناسب هو الذي شجعني لاختيار الموضوع الذي عكفت على دراسته. وفي ذلك الحين نصحني سلام قائلا:«انك مقدم على دراسة وبحث موضوع يتطلب منك الكثير من التأمل والتفكير، فان كنت قد اقدمت عليه فما عليك الا أن تبحثه بعمق وان لا تكتفي بنظرية الكتلة الصغيرة التي عرفتها انا وعلمتك اياها». فعليه غصت في رياضيات بنظرية الكتلة (gruop) وهي الآلية التي بواسطتها يمكن دراسة نظرية التناسق. وكانت النتيجة من دراستي وبحثي التعريف بنماذج ابرزتها المئات من ذرات الجسيمات المختلفة والتي قد اكتشف العديد منها للتو. وفي السنة ١٩٦٢ بدأت أدرك المبنى الذي تقوم عليه هذه الجسيمات والتي كانت جمعيها

مزيجا من ثلاث طوبات أساسية أطلق عليها فيما بعد اسم الكواركس (Quarks). وقد تقين سلام من تقريري هذا حول نظرية الكتلة.

وفي حوالي سنة ١٩٦٨ أتم سلام اقتحامه الهام وأصبح يهتم بتحطيم فكرة التناسق وكان قد سبق ونشر مساهمته الأولى في هذا المجال سنة ١٩٦٢ بالاشتراك مع جولدستون وواينبرج.

في ذلك الوقت أمسك سلام الثور من قرنيه وعرض في اجتماع دولي عقد في جوتيبورج صيف ١٩٦٨ نظريته المتكاملة عن توحيد تفاعل القوة الضعيفة مع القوة الكهرومغناطيسية والتي سميت منذ ذلك الحين بالقوة الكهروضعيفة Electroweak. وهكذا استعمل سلام مفهوم التناسق عند جلاشو واليه اكتساب الحجم عند هيجز- كيبل مضيفًا اليها تصوره ونظرته الفيزيائية وقدم نظرية كاملة كأول مثال على توحيد القوى منذ قدم ماكسويل نظريته حول توحيد الكهرومغناطيس والضوء قبل مئة سنة. هذه النتيجة نفسها عرضها واينبرج ايضا ولكن بصورة مستقلة عن سلام، وبذلك تحققت اول تنبؤات عن النظرية الجديدة سنة ١٩٧٨ وعليه منح سلام جائزة نوبل سنة ١٩٧٩ وشاركه في هذه الجائزة كل من واينبرج وجلاشو. اما الوسائط الهائلة لتفاعل القوة الضعيفة فقد اكتشفت سنة ١٩٨٢، ومن هنا مضى سلام يقدم نظريات جديدة وتنبؤات أخرى خلال السبعينات وعلى الخصوص مشروعه لتوحيد القوة الكهروضعيفة مع التفاعل الذري القوي. كذلك اقترح أن هذه القوة يمكن أن يكون لها مركب شبيه بمركب الجاذبية القوية، وهذا ما يبقى ينتظر بحثا على المستوى النظري التجريبي متواصلا خلال السنوات القادمة.

ان مساهمة سلام في مجال علوم الفيزياء هي وجه واحد فقط من أرجه نشاطه العلمي، ولأنه كان يعي أهمية العلوم في مجال التطور الاقتصادي والإجتماعي والثقافي فقد بادر الى إنشاء المركز الدولي للفيزياء النظرية في تريست سنة ١٩٦٢ تحت اشراف اليونسكو والهيئة الدولية للطاقة الذرية IAEA وقد أصبح هذا المركز المدرسة الاولى التي وجدت فيها الدول النامية وسيلة لتدريب خيرة علمائها وليطلعوا عن كثب على أكبر البرامج العلمية المتطورة.

ومن الأمور الأخرى التي بادر اليها سلام في هذا المجال انشاء اكاديمية العلوم للعالم الثالث.

كذلك فانه ضحى بالكثير من قدراته من أجل تطوير بلده الباكستان علميا على الرغم من المصاعب التي كان يواجهها.

لقد كان سلام أحد العلماء الرائدين في مجال فيزياء الجسيمات والفيزياء الميدانية في النصف الثاني لهذا القرن وان تراثه للقرن الواحد والعشرين لهو درس حول الدور الأساسي للتناسق والتجانس.

Salam the Humanitarian

ABDUS SALAM AND PAKISTAN

MALEEHA LODHI
Pakistan High Commissioner, UK

It is humbling to speak about Dr. Salam — the only Pakistani conferred with the Nobel Prize — the man who put Pakistan on the map of international science.

Dr. Salam is perhaps the most accomplished scientist who spent many years of his active career at Imperial College. I understand that 2007 marks the centenary of Imperial College. It is also the 50th anniversary of Dr. Salam's association with the College. It is, therefore, entirely appropriate that we have gathered here to pay tribute to Dr. Salam.

Dr. Salam not only excelled in Theoretical Physics but was also the epitome of human dignity. He was a great believer that scientific thought is the common heritage of mankind and that developing nations needed to help themselves and invest in their own scientists. In fact he was one of the main architects and pioneers of modern science in Pakistan. His advocacy of developing a scientific base for the country was truly visionary — a vision we are trying to translate into action today.

His active association with the development of science and technology in Pakistan dates back to 1955 when he was included in the Pakistan delegation for participation in the first UN conference on Peaceful Uses of Atomic Energy in Geneva. The development of peaceful uses of nuclear energy power in Pakistan was a case Dr. Salam made vigorously and with deep commitment. By the late 50s, Professor Salam's important contribution to Elementary Particles Physics was already acknowledged in Pakistan. Punjab University conferred an honorary degree of DSc on him in December 1957.

Ayub Khan, the then President of Pakistan, paid tribute to Dr. Salam in the following words: "His attainment in the field of science at such a young age is a source of pride and inspiration for us". As chief scientific adviser to the President from 1961 to 1974, Dr. Salam was the moving spirit behind the establishment of a number of research institutes in the country. He was a member of PAEC, a member of the Scientific Commission of Pakistan, and Founder Chairman of SUPARCO, the Space and Upper Atmosphere Research Commission.

In 1979, the Government of Pakistan conferred the highest civilian award, the Nishan-i-Imtiaz on Dr. Salam for his outstanding contribution in science.

Dr. Salam will always be remembered for his humility and humanity. As we pay tribute to him, let me say Pakistan is proud of him, his achievements, but above all, his humanity.

Donal Bradley chatting with Graziela De Nadai-Sowrey.

All photographs reprinted courtesy of Meilin Sancho and Blackett Lab.

Umar Salam.

Faheem Hussain and Michael Duff.

Dr. Maleeha Lodhi, Pakistan High Commissioner, UK and Michael Duff.

The Salam grandchildren.

Nobel Laureate Gerard 't Hooft chatting with John Charap and Chris Pope.

Chris Isham.

David Weir receiving the Abdus Salam Imperial College Physics Graduate prize from Mirza Fakhar Ahmad Sb., President of Ahmadiyya Muslim Youth Association, UK.

Tom Kibble, Rashid Salam and Louise Johnson.

Tom Kibble, Ahmad Salam (holding the Nobel Medal) and Rashid Salam.

Michael Duff introducing Gordon Fraser.

Kelly Stelle and Chris Isham.

THE INTERNATIONAL CENTRE FOR THEORETICAL PHYSICS

FAHEEM HUSSAIN

School of Science and Engineering,
Lahore University of Management Sciences,
Sector U, DHA, Lahore 54792, Pakistan
fhussain@lums.edu.pk
hussainf@ictp.it

This talk traces in brief the genesis of the Abdus Salam International Centre for Theoretical Physics, Trieste, as one of Prof. Abdus Salam's major achievements. It outlines why Salam felt the necessity for establishing such a centre to help physicists in the developing world. It situates the founding of the Centre within Salam's broader vision of the causes of underdevelopment and of science as an engine for scientific, technological, economic and social development. The talk reviews the successes and failures of the ICTP and gives a brief overall view of the current status of the Centre.

Sir Richard Sykes, Your Excellency the High Commissioner Maleeha Lodhi, ladies and gentlemen,

I would like to begin by thanking the organisers, in particular, Prof. Mike Duff, for having invited me to speak at this meeting commemorating the 50th anniversary of Prof. Abdus Salam's move from Cambridge to Imperial. It is indeed a great honour to be here in this August company and it is a special pleasure for me to be back at my old college, which is also celebrating its 100th anniversary this year.

Fifty years ago when Salam was moving from Cambridge, to establish and build the Department of Theoretical Physics at Imperial, his thoughts were already fixed beyond to another dream, to another aim. Salam had gone up to St. John's College, Cambridge in 1946 as a Foundation Scholar from where he obtained a double first in Mathematics (Wrangler) and Physics. He went on to complete his Ph.D. in Theoretical Physics from the Cavendish Laboratory in 1952. He returned to Lahore in 1951 to teach mathematics at Government College, Lahore. He had come back to Pakistan with the intention of founding a school of research, but it soon became clear that this was impossible. He found that he was completed isolated from his peers and there was no chance for him to pursue his love of physics. If it had not been for this we would have been celebrating in Lahore rather than in London. He was compelled to leave Lahore and join St. John's at Cambridge in the early 50s so that he could continue to do research in physics. But his experience in Lahore marked him for the rest of his life. This is what he had to say of this period:

> "Looking back on my own period of work in Lahore, as I said, I felt terribly isolated. If at that time someone had said to me, we shall give you the opportunity every year to travel to an active centre in Europe or the United States for three months of your vacation to work with your peers; would you then be happy to stay the remaining nine months at Lahore?, I would have said yes. No one made the offer. I felt then and I feel now that this is one way of halting the brain drain, of keeping active men happy and contented within their own countries. They must be kept there to build for the future, but their scientific integrity must also be preserved. By providing them with this guaranteed opportunity for remaining in contact with their peers, we believe we are making a contribution to solving the problem of isolation."[1]

This was the genesis of the International Centre for Theoretical Physics, which is now named after Salam. But why should have Salam worried so much about science in the developing countries? Here we come to his other passion. He was after all from a poor country and he saw the immense and growing gap between these countries and the developed countries of the West and he came to the conclusion that science was the key to development. He was angry about the poverty and underdevelopment of the poor countries.

Salam was convinced that the most important factor that condemns the majority of the world to live in misery and suffering was the lack of scientific development. He said in 1988:

> "This globe of ours is inhabited by two distinct types of humans. According to the UNDP count of 1983, one-quarter of mankind, some 1.1 billion people are developed. They inhabit two-fifths of the land area of the earth and control 80% of the world's natural resources, while 3.6 billion developing humans — 'Les Misérables' — the 'mustazeffin' — live on the remaining three-fifths of the globe. What distinguishes one type of human from the other is the ambition, the power, the élan which basically stems from *their differing mastery and utilisation of present day Science and Technology.* It is a *political* decision on the part of those (principally from the South) who decide on the destiny of developing humanity if they will take steps to let Les Misérables (the 'mustazeffin') *create, master*, and *utilise modern Science and Technology.*"[2]

(By the way, those of you, who knew Salam, will remember that he had a complete mastery over the English language. In fact at one point he had even thought of reading English at Cambridge.)

I believe that these words are even truer today and they are tragically borne out by the events of the last few years. The future of the earth and the very existence of our civilisation depend on whether we, collectively, are able to eliminate poverty and inequality from this world of ours. Of course there are other causes, like the inheritance of colonialism, the existence of neo-colonialism, the return of unabashed imperialism, superstition, the venality of our rulers but I believe that Salam was right in identifying the lack of science as one of the crucial factors which contributes to our backwardness.

In the same article Salam quotes Alfred North Whitehead:

> "In the conditions of modern life, the rule is absolute: the race which does not value trained intelligence is doomed .... Today we maintain ourselves, tomorrow science will have moved over yet one more step and there will be no appeal from the judgement which will be pronounced ... on the uneducated."[2]

Further, Salam wrote in the same paper

> "Today the Third World is only slowly waking up to the realisation that in the final analysis, *creation, mastery* and *utilisation of modern Science and Technology* is basically what distinguishes the South from the North. On Science and Technology depend the standard of living of a nation. The widening gap in Economics and influence between the nations of the South and the North is essentially the Science and Technology gap. *Nothing else* — neither differing cultural values, nor differing perceptions or religious thoughts, nor differing systems of economics or of governance — can explain why the North (to the exclusion of the South) can master this globe of ours and beyond."[2]

He therefore felt that developing countries needed to train scientists quickly but at the same time they needed to create conditions to retain them in their own countries. From his own experience he was painfully aware of the brain drain.

As for training scientists, already at Imperial, he made a concerted effort to recruit the best and brightest students from the Third World. Those of you who were here in the 60s may recall the large number of students from Africa and Asia in the Diploma programme, which was then a prerequisite to be admitted to the Ph.D. programme. In fact his student, Dan Akyeampong, from Ghana was the first African to obtain a Ph.D. in theoretical physics. I hope this tradition is still maintained at Imperial today.

Salam was driven by this reality and also by his own personal experiences on returning to Lahore after his studies at Cambridge to do something about it. He realised that one of the ways to help scientific research in developing countries would be to encourage scientists from developing countries to stay in their own countries while still being able to pursue their research interests and without being cut-off from the larger scientific community. In the late fifties he thus conceived of an international centre for physics geared to helping physicists in the developing countries. This centre would provide a place where physicists from developing countries could visit periodically and where they could interact with leading international physicists and where they could pursue their own research interests before returning to their own countries. He envisaged this as a centre where research would be done at the frontiers of science and not any second-rate research.

However Salam did not just have a utilitarian view about the benefits of science. The other factor driving Salam's vision was that he believed passionately that science was the common heritage of mankind. He understood this to mean not only that all mankind should benefit from the fruits of science but also that all past civilisations, Egyptian, Greek, Chinese, Indian, Islamic, etc., have made significant contributions to science and technology. He wanted poor countries to be not just consumers of the products of science and technology but also creators of science. In the quote above he stresses creation and mastery and not just utilisation. He could not understand why somebody from the remotest part of Africa and Asia should not be able to understand and contribute at the highest level in scientific research, for example cosmology and particle physics, irrespective of whether what he was doing was immediately relevant for economic development. In this he differed from much of the policies followed in developing countries where the emphasis is on technology and applied research. He understood very well that you couldn't have high technology without high science.

The idea that science is the common heritage of mankind also precludes the notion that science is Western science. For Salam science was universal and there could be no such entities as Third World science or Islamic science or Marxist science.

Salam was nothing if not a dreamer. Any other young man, remember that he was only in his thirties at that time, would have been daunted with the task. But Salam had another quality, which distinguished him from ordinary scientists. He was a consummate politician and a charismatic person. Also he was lucky as he was part of the delegation from Pakistan to the General Conference of the International Atomic Energy Agency, in recognition of his scientific achievements at that young age. Salam, this young bold man, took this opportunity to make the audacious proposal to establish an international centre for physics at the IVth regular session of the General Conference of the IAEA in Vienna on 22 September 1960. This preposterous proposal was nearly laughed out of court. His proposal was opposed by the delegates of many of the developed countries, in particular the United States. There were comments like: Why do underdeveloped countries need advanced science? Engineering and technology is good enough for them. And some really pejorative comments were made like: "A centre for physics for underdeveloped countries will produce underdeveloped physics." But Salam was able to overcome these objections because he used his wide contacts and his scientific reputation in the international physics community. He had these wide contacts in 1960 because already at this young age he was known as a brilliant scientist who had done fundamental work on renormalisation, parity violation, etc. Remember that this was before his work which won him the Nobel Prize. He mobilised these contacts to convince the major countries to support the idea. The Board of Governors of the IAEA finally approved the establishment of the ICTP in February 1963 and the selection of Trieste as its seat in June 1963. The choice of Trieste owes a lot to the activity of Paolo Budinich in Trieste and Rome. The ICTP was officially inaugurated at Trieste, Italy, on 5 October 1964 under the aegis of the IAEA.

In the words of Alexis De Greiff, another Imperial student who did his Ph.D. from Imperial's History of Science Department, the successful establishment of the ICTP was one of the successes of Salam as

> "a scientific diplomat, a representative of different communities acting in different social and political settings. He was a diplomat who represented the third world before the audience of industrialised countries; at the same time, he represented the international scientific community and western progress in Pakistan (and in the third world). His own life seemed to embody, indeed to represent these diverse worlds. As a scientist he carried great authority before politicians in Pakistan (and in the Third World), and as a Muslim born in poor Pakistan, he seemed to be naturally invested with the authority to speak on behalf of the third world. His double identity as a third world citizen and as a professor at a British university made of him a cultural amphibian, "a man of two worlds": the world of physics and the world of the politics for development, the first and the third world."[3]

This aspect of Salam was embodied in the ICTP. Salam was naturally attracted to the UN system as he viewed it as an instrument to promote the development of the Third World. Salam got into the UN system through the Atoms for Peace Conferences where he was appointed as secretary at the Geneva Conference in 1955 and again in 1958. Salam was elected member of the board of governors at the IAEA for 1962–63. It is worth noting that he got access to this UN technical agency through his own government which nominated him as a member of the Pakistani delegation. Between 1964 and 1975 he was a member of the United Nations Advisory Committee for Science and Technology and from 1970 to 1973 he was a member of the UN Panel and Foundation Committee for the United Nations University. This list is not exhaustive, yet it illustrates his engagement in the diplomatic and political worlds of science and development.

Thus his dream was realised in 1964 by the establishment of the International Centre for Theoretical Physics in Trieste under the aegis of the IAEA in Vienna. I will not go further into the history of the establishment of the centre except to mention that this centre was established over the objections of some countries. By the way the US still looks with suspicion at the activities of the Centre. You may wonder why the Centre was established at Trieste. In 1963 there were several possible choices of location for the Centre apart from Trieste. The choice of Trieste is a story of historical conjunction. Salam had been invited to a symposium on particle physics in June 1960 and had developed close contacts with the local physics community and in particular Paolo Budinich, a young Italian who had worked with Werner Heisenberg after the war. Budinich wanted to establish physics in Trieste and he realised that if he could get the new centre to Trieste it would give a great boost to the local physics community. Budinich, like Salam, is an excellent politician. He quickly mobilised the local authorities and through them the Italian government to make the offer of Trieste as the place where the Centre should be established. I am still amazed how he could get the notoriously slow Italian bureaucracy to move so fast! In 1960 Salam had met the Prince of Torre e Tasso (a branch of the Bavarian Thurn und Taxis family). The Prince offered a piece of land next to the Miramare Park where the campus of the future ICTP would be constructed.

Trieste is in the far northeastern corner of Italy and had been disputed territory, between Yugoslavia and Italy, after the war until the mid-fifties when it was finally given to Italy. The Italian authorities felt that bringing an international research centre to this remote corner of Italy would boost the local economy, particularly so as the Trieste port was cutoff from its natural hinterland which was then part of Yugoslavia and they were farsighted enough to see that the Centre would become a pole of attraction for the establishment of other research institutions. For Salam, Trieste was attractive as it did not have the distractions of a big city, so that scientists could concentrate on their work, and was still fairly accessible.

The ICTP started off with a handful of people temporarily located in a local government building, which now houses the offices of the Region of Friuli-Venezia-Giulia, in downtown Trieste. Ray Rivers was one of the early students to go to Trieste with Salam. I was not there but I am told that the atmosphere was very relaxed and informal with Salam and his students going to dinner together at the railway mensa (cafeteria) in the evenings to continue their physics discussions sharing simple proletarian fare with railway workers. Now the ICTP is a huge endeavour. Here Salam instituted the famous "Associateships" which allowed deserving young physicists to spend their vacations at the ICTP in an invigorating atmosphere, in close touch with their peers in research and with the leaders in their own field, losing their sense of isolation and returning to their own country for nine months of the academic year refreshed and recharged.

From the beginning the Centre has been supported mainly by Italian funds. The Italian government has been very generous in this regard. Even now most of the funds come from the Italian government under an Italian law. From the beginning the Centre was run under a tripartite agreement between the IAEA, UNESCO and the Italian government. From its establishment till the early 1990s the Centre was under the administrative control of the IAEA in Vienna. From the early 1990s the Centre passed under the control of UNESCO in Paris although the major part of the funds are still Italian.

What has been the impact of the ICTP? Here let me me again quote Alexis De Greiff:

> "The ICTP became the most important institution for training and research for Third World physicists and an emblematic centre for international collaboration. Its impact has not been measured, but the centre

makes an estimate that, according to my own work, is reasonable: it has been visited by at least one physicist of every institution devoted to theoretical physics in the Third World. Statistics carried out by the centre indicate that the number of visitors has risen to 60,000 people from 150 countries since 1965. It was conceived as a space for international collaboration between scientists from the industrialised and the Third World countries."[3]

In fact recent statistics complied by the Centre show that the total number of visitors has risen to over 100,000 since 1964 with more than 5000 visitors every year.

Further on De Greiff says:

> "The ICTP has been presented as an alternative offered to scientists who are ready to emigrate to industrialised countries for the lack of stimulus and support in their own countries. He (Salam) viewed the ICTP as a new Toledo, or a new Academy of Gondeshapur and identified Third World scientists with the story of Michael the Scot:
>
>> 'Seven hundred and fifty years ago, an impoverished Scotsman left his native glens to travel south to Toledo in Spain. His name was Michael, his goal to live and work at the Arab universities of Toledo and Cordova .... His interests lay in the sciences of astrology and alchemy, then fashionable in Scotland. But once in Toledo, Michael formed the ambitious project of introducing Aristotle to Latin Europe, translating not from the original Greek, which he did not know, but from the Arabic translation taught in Spain [$\cdots$] Toledo's school, representing as it did the finest synthesis of Arabic, Greek, Latin and Hebrew scholarship, was one of the most memorable of international essays into international collaboration.'[4]
>>
>> ....
>
> In Salam's cyclic theory of history, the present situation was symmetrical, but the roles had been inverted: it was now the new Michaels from the south who had to come to the north to learn science. The importance was that they should return to their countries. For that the ICTP created a mechanism called the 'associate scheme' in which scientists who work in the Third World visit the centre for periods from one to three months during five years. In Salam's words 'the idea was that a deserving young man may spend his period of vacation in an invigorating environment ... and charge his batteries with new ideas, still spending the remaining nine months of his academic year back at home working in his own university.'"[3]

Here are some facts about the current state of the Associates Programme. It is the oldest programme at the Centre and has currently 800 appointed Associates.

Past Associates have gone on to hold distinguished positions in their own countries, amongst these are 1 President of a Republic; 11 Ministers or Deputy Ministers; 2 Members of Parliament; 7 Advisors to Presidents or Prime Ministers; 18 University Presidents or Vice Chancellors; 22 Deans of Faculty; about 150 Heads of Laboratory or Departments.

There is another aspect of the ICTP, which is sometimes forgotten. It's a place where scientists from countries which are hostile to each other can come and can do science together. During the Cold War, Soviet scientists were isolated and it was difficult if not impossible for them to meet and interact with their Western colleagues. Trieste was one of the few places where this was possible, where Soviet scientists could meet their American colleagues, and it played a significant role in ending the isolation of Eastern bloc physicists. This is one of the few places where physicists from Pakistan and India, for example, can and have worked together. In this sense the Centre has played a very positive role in fostering the idea that science is a truly international activity, which transcends borders and ideologies. Young people visiting the Centre make long lasting friendships and form collaborations with their peers from around the world irrespective of different cultures. In this sense the ICTP represents the best of the United Nations system.

As you can see from the statistics the ICTP developed into not just a centre for Third World scientists but it also became an active research centre in its own right doing physics at the highest level led by the example of Salam himself. Because of him the best minds from Europe and the States were attracted to the Centre to carry on their own research and to mentor younger colleagues from developing countries. Many of the conferences and schools were intended mainly for students from developed countries, e.g. the best string school in Europe is the annual spring school in Trieste, which is attended by students from the best universities in Europe.

I believe that the Centre has been a success but it should still be looked at as a drop in the ocean. It has had limited success in spite of the huge statistics which I have shown you. Definitely it has helped individual scientists to maintain their scientific integrity but Salam had high hopes in the 60s and 70s about how the Centre could really kick start physics in the developing world. However this did not happen except in a few countries. Its activities have been very helpful in the development of physics in China, India, Brazil and Argentina and some other countries. This is because the abovenamed countries were able to use the Centre. But most of the Third World is still very, very backward in the sciences. My own country is extremely backward in science because we have had a succession of governments, which were not interested in science and education. The level of physics in Pakistan is abysmal. The problem is that the ICTP cannot do anything from the outside. As usual change is possible only from the inside and if this is not forthcoming then no progress is possible. The role of the ICTP is basically as a catalyst and as a place, which provides the facilities if anyone is interested in using them.

Salam had envisaged the Centre as a prototype of many other such institutions in other fields spread out in the developing countries. Unfortunately this did not materialise except for the International Centre for Genetic Engineering and Biology based in Trieste and New Delhi.

However the Centre was a great success as a catalyst for the establishment of scientific institutions in and around Trieste. The International School for Advanced Studies (Scuola Internazionale Superiore di Studi Avanzati to give its full Italian title), a part of the Italian university system, grew out of efforts from the Centre. The Italian light source, the synchrotron Elettra, was built on the plateau above Trieste and now around this there is a flourishing science park. All these institutions are now known collectively as the Trieste system and are the pride of the Italian state. Salam paid back Trieste handsomely for the faith the city put in him when establishing the ICTP.

People at the Centre realise that the severest problems lie in Africa, which has been completely bypassed in the era of globalisation and neo-colonialism. Most of the Asian countries, excluding countries like Pakistan and Afghanistan and others, and Latin America now have a growing scientific endeavour but the continent which is missing is Africa. More and more of the activities of the Centre are now quite rightly geared towards Africa.

I did not have time to talk about Salam and Pakistan. It is a matter of record that Pakistan did not treat Salam well. He was not honoured in his own country for reasons of religious intolerance. There are no streets named after him. He is not mentioned in Pakistani school books as a hero to emulate whereas much lesser men are talked about. One of the dreams of Salam was to develop Pakistani science and physics in particular. I am afraid that this dream was not realised. His advice was consistently ignored. Pakistan is still very backward in science. However recently there have been attempts to redress both these issues. The present government is aware of the backwardness of Pakistani science and university education and is putting a lot of cash into higher education and science to make up for the years of neglect.

Although the powers that be in Pakistan are still wary of honouring Salam openly, nevertheless there have been some small efforts by individual institutions. A few years ago Government College University, Lahore instituted a Salam Chair; in fact some years before Imperial. At the recently concluded Nathiagali Summer College on Contemporary Physics, Prof. Atta-ur-Rahman, Chairman of the Higher Education Commission proposed that the National Centre for Physics should be named after Salam. I doubt if this will happen given the reluctance and opposition of some people within the Pakistani scientific establishment.

But what I want to announce here is the institution of a new endowed Salam Chair at the Lahore University of Management Sciences, a private university. This was originally a business school but now the industrialists and businessmen behind this university have become aware of the necessity of science and technology. So they are funding a huge expansion in terms of setting up a School of Science and

Engineering which will start its teaching programme in 2008. They have decided to give a boost to physics by setting up an endowed Chair named after Prof. Salam. This will be the fourth Salam Chair after Government College, Lahore, the Abdus Salam ICTP and Imperial.

Acknowledgments

I would like to thank the Abdus Salam International Centre for Theoretical Physics for providing me the opportunity to work there for many years. Here I imbibed the internationalist spirit embodied by Salam. I had the privilege to work with Salam and I acknowledge with gratitude the strong influence he had on my life and work. I would also like to thank Prof. K.R. Sreenivasan for providing me with the statistics which I used to illustrate this talk.

References

1. Abdus Salam, The Isolation of the Scientist in Developing Countries, *Minerva* **IV** (4), (1966), reprinted in *Ideals and Realities. Selected Essays of Abdus Salam* (Third Edition), edited by C. H. Lai and Azim Kidwai (World Scientific, Singapore, 1989), pp. 211–216.
2. Abdus Salam, *Notes on Science, Technology and Science Education in the Development of the South* (prepared for the 4th Meeting of the South Commission, 10–12 December 1988, Kuwait), *Ideals and Realities. Selected Essays of Abdus Salam* (Third Edition), edited by C. H. Lai and Azim Kidwai (World Scientific, Singapore, 1989), pp. 55–183.
3. Alexis De Greiff, Abdus Salam: A Migrant Scientist in Post-imperial Times, *Economic and Political Weekly* **41** (3), 228–234 (2006).
4. Abdus Salam, The United Nations and the International World of Physics, *Bulletin of Atomic Scientists*, February 1968, pp. 14–15.

A PERSONAL ACCOUNT OF PROFESSOR ABDUS SALAM

K. R. SREENIVASAN
Abdus Salam Research Professor
Director, ICTP, Trieste, Italy

1. Preamble

Professor Abdus Salam makes a great copy for a biography. He was born into a relatively poor family but made good in incredible ways; he was from a poor Asian country with no strong traditions of doing modern science but broke through those barriers by doing great physics and attaining the pinnacle of glory in the Western world; he was the conscience of good science in all developing countries; he felt deeply for the lost glory of the Islamic world and agonized over it with untold burden; he created such fine institutions as ICTP and TWAS and provided them with inspiring leadership; he was admired by scientists all over the world. He was extraordinarily colorful as a scientist and a human being and an intellectual of the first order. He lived like a prince and died in agony. In short, he was many wonderful things — all in one lifetime.

Yet, controversy did not escape Salam in his own lifetime and later; his vision could be grand but details sloppy; he was at ease with generals as much as he was with Nobel Laureates; he was at once humble and vain; even as he cajoled his country into getting better in science, he was hindered at the official level from achieving his powerful desire to become the Director-General of UNESCO; he felt the compelling need to proclaim his religiosity even as he was formally banned from the fold; he was accused of taking ambiguous stand on atomic weapons; and so forth.

It is thus not a surprise that there is an increasing interest in Salam's rich and varied life and accomplishments. The troubled times of today are especially appropriate for taking this interest because here was a person who professed unabashedly to drawing inspiration from Koran but demonstrated without an iota of doubt that this connection did not interfere with his scientific prowess — indeed, in his mind, his religious outlook added to the richness of his science. Here was a person who succeeded more than once, in more ways than one, and against many odds.

Several people have thus recently expressed interest in writing about Salam or making documentaries of his life. This renewed interest has led to requests from potential authors, both secular and religious, for access to Salam's files and to people who knew him well. I have made the catalogued material readily available to all as a matter of principle, but must admit to a misgiving towards opening his personal files to anyone. My one principal reason is that very few people have demonstrated the depth of perception to view Salam's complex personality sympathetically and in totality, without depriving the posterity of the contradictions of his life and without defying him, and yet to draw critical lessons from it. My belief is that Salam, who was clearly aware of his special status, would have wanted nothing less. There is no progress otherwise.

An authoritative biographical memoir of Salam is thus overdue. Salam is one of my heroes and I am pleased to write about him for these Proceedings, but this article is not even remotely at the level that is needed. While sitting in what used to be once his office, I cannot help thinking the thoughts that he must have entertained, the kinds of ambiguities he willingly embraced — without, however, sacrificing clarity in his own personal science. His task must have required a great deal of tact, optimism, sympathy, understanding, opportunism, high-level persuasions and low-level browbeating — and endless frustrations. That he excelled in this self-appointed role is for me a matter of great admiration and pride in the totality of Salam.

In the following sections, I shall briefly attempt to record my impressions of Salam. The part of the text is somewhat reworked from an article prepared for the commemoration of the 10th anniversary of Professor Salam's death.

2. Abdus Salam in His Own Words

It is best to let Salam speak in his own words on how he evolved into a great physicist from modest beginnings. I reproduce below a brief article that was culled a few years ago from some of Salam's writings.[1] The article also speaks directly to his vivacious and engaging personality.

> "I was born in the country town of Jhang, then part of British India, now Pakistan, in 1926. My father was a teacher and educational official in the Department of Education and my mother was a housewife. I had six brothers and one sister. My family was by no means rich. My father took a vast amount of interest in my school work. He had great ambitions for me. I was destined for the Indian Civil Service, entry to which was by competitive examination. However, this was not to be — as events in my life took a different turn.
>
> When I was at school in about 1936 I remember the teacher giving us a lecture on the basic forces in Nature. He began with gravity. Of course we had all heard of gravity. Then he went on to say "Electricity. Now there is

a force called electricity, but it doesn't live in our town Jhang, it lives in the capital town of Lahore, 100 miles to the east". He had just heard of the nuclear force and said "that only exists in Europe". This is to demonstrate what it was like to be taught in a developing country.

When I was 14, I won a scholarship to Government College, Lahore, with the highest marks ever recorded. I recall that when I cycled home from Lahore, the whole town turned out to welcome me. I wrote my first research paper when I was about sixteen years of age. It was published in a mathematics journal but I wasn't actually hooked on research till I went to Cambridge University.

I was very fortunate to get a scholarship to go to Cambridge. The famous Indian Civil Service examinations had been suspended because of the war and there was a fund of money that had been collected by the Prime Minister of Punjab. This money had been intended for use during the war, but there was some of it left unused and five scholarships were created for study abroad. It was 1946 and I managed to get a place in one of the boats that were full with British families who were leaving before Indian Independence. If I had not gone that year, I wouldn't have been able to go to Cambridge; in the following year there was the partition between India and Pakistan and the scholarships simply disappeared.

At Cambridge, I achieved a First in the Mathematics Tripos in two years. I still had a third year free in the sense that I had the scholarship and the choice of whether to go on with higher mathematics — that's part III of the mathematics tripos — or to do the physics tripos. On the advice of my tutor, Fred Hoyle, who said "If you want to become a physicist, even a theoretical physicist, you must do the experimental course at the Cavendish. Otherwise, you will never be able to look an experimental physicist in the eye". I joined the Cavendish Laboratory where Rutherford had carried out his experiments on the structure of the atom. The Cavendish was an outstanding laboratory for experimental work and a focus for physicists around the world. However, I had very little patience with experimental equipment. To be a good experimenter you must have patience towards things which are not always in your control. I think a theoretician has got to be patient too, but that is with something of his own creation, his own constructs, his own stupidities.

The very first experiment I was asked to do was to measure the difference in wavelength of the two sodium D lines, the most prominent lines in the sodium spectrum. I reckoned that if I drew a straight line on the graph paper then its intercept would give me the required quantity I wanted to measure. Mathematically, a straight line is defined by two points and if you take one other reading then mathematically that should be enough since you then have three points on that line, two to define the straight line and the third one to confirm it. I spent three days in setting up that equipment.

After that I took three readings and took them to be marked. In those days the marking of experimental work in the class counted towards your final examination. Sir Denys Wilkinson was one of the men who supervised our experimental work, and I took it to him. He looked at my straight line, and asked "What's your background?" I said "Mathematics". He said "Ah, I thought so. You realise that instead of three readings you should have taken one thousand readings and drawn a straight line through them". I had by that time dismantled my stuff and didn't want to go back. So I tried very hard to avoid Denys Wilkinson during the rest of the year. I still remember when the results came out in 1949. I was looking at the results sheets hung in the Cavendish and Wilkinson came up behind me. He looked at me and said "What sort of class have you got?" and I very modestly said "Well, I've got a first class". He turned full circle on his heel, three hundred and sixty degrees, turned completely round, and said "Shows you how wrong you can be about people".

I went back to Lahore in 1951 and taught there at the University. But as a physicist, I was completely isolated. It was very difficult to get the journals and keep in touch with my subject. I had to leave my country to remain a physicist. Now, it is the lack of this contact with others that is the biggest curse of being a scientist in a developing country. You simply do not have the funds, the opportunities, which those from richer countries enjoy as a matter of course. There are not the communities of people thinking and working in the same fields. This is what we have tried to cure by bringing people together at the International Centre for Theoretical Physics which I founded in Trieste in 1964. The Centre provides the possibility for scientists to remain in their own country for the bulk of the time, but come to the Centre to carry out research for three months or so. They meet people working in the same subject, learn new ideas and can return to their own country charged with a mission to change the image of science and technology in their own country.

I returned to Cambridge in 1954 as a lecturer and Fellow of St. John's College. Three years later, I accepted a professorship at Imperial College, London, where I succeeded in establishing one of the best theoretical physics groups in the world.

The pinnacle of my physics career came in 1979 when I shared the Nobel Physics Prize with Sheldon Glashow and Steven Weinberg for our unification of electromagnetism and the weak nuclear force in the "electroweak" (a word which I invented in 1978) theory, one of the major achievements of twentieth-century physics. This theory had made predictions that could be verified by experiment. The most revealing of these was that a new particle exists at extreme energies. To test this theory we had to convince the experimental physicists working on the great particle accelerators to build new equipment: To create, in principle, conditions that would be similar to those first few moments in the birth of the universe. In 1983 the final confirmation was obtained with the discovery that the predicted particles — the intermediate vector bosons — did exist. Called W^+, W^- and Z^0, these hypothetical particles were seen for a few fleeting moments under the cosmic conditions of the CERN accelerator. This temporary existence was enough to demonstrate that the unification theory was an accurate description of the fundamental nature of matter. This experimental verification led to the award of the Nobel Prize to Carlo Rubbia and Simon van der Meer in 1984."

I might add the following postscript: Salam held his professorial position at the Imperial College from 1957 until 1993 with distinction. From 1964 until 1993, he was concurrently the Director of the International Centre for Theoretical Physics (ICTP), where he provided both the physical drive and the lofty vision. For a period of time, he played various advisory roles for the government of Pakistan, and acted as a spokesperson for science in developing countries, especially Islamic countries. Salam fell prey, around 1985, to a neural disorder known as Progressive Supranuclear Palsy, involving the death of selected neurons in the brain. Yet, he exerted himself greatly to carry on his responsibilities for several more years at ICTP (and also TWAS). Those who knew him remember them as his difficult years. He passed away at his home in Oxford on 21 November 1996.

3. Brief Remarks on Salam's Physics

Salam's place in physics is described in several places,[2] but it is useful to understand it in his own words. In an undated popular talk given sometime after 1979, he described his work as a major milestone in the quest for unification of forces of nature. He first described Newton's role in the unification of celestial and terrestrial gravitation — an idea that is now commonplace, but undoubtedly revolutionary at the time. Then Einstein's theory of relativity defined gravitation through the curvature of the space-time manifold. Space and time were never again to be considered in separate terms.

On another branch of unification, Faraday realized that electricity and magnetism were two aspects of the same physical phenomenon, and Maxwell wrote down his beautiful equations describing the theory of electromagnetic radiation.

The next set of forces deals with nuclear structure. The weak force is the second weakest after gravity, responsible for radioactive decay and neutrino interactions. Enrico Fermi understood the basics of weak interactions while studying the decay of radiation. The weak force occurs in the decay of nuclear particles requiring, as learnt later, a change of a quark of one flavor to another. The theory that describes the unified electromagnetic and weak interactions is the Standard Electroweak theory, which, in large part, is the work of Sheldon Glashow, Abdus Salam himself and Steven Weinberg, for which they shared the 1979 Nobel Prize.

The strong force is short-ranged, acting over ranges of order 10^{-13} cm and is responsible for holding together the nuclei of atoms. It is important for both nuclear fission and fusion. Despite existing gaps, there is strong evidence to suggest that a theory that unifies strong forces with electroweak forces is required to make sense of the Universe. This is no place for a historical survey of developments in this quest (nor am I qualified for the task), but it suffices to say that, along with Jogesh Pati (one of ICTP's Dirac Medalists), Salam played an important role in the development of this part of physics as well (see the collection of papers in Ref. 2).

The quest for unifying all forces including gravity has been the focus of attention in high-energy physics. It is not surprising that Salam took interest in unifying gravitational field, and again I refer to Ref. 2 for some of his papers with his long-time collaborator, John Strathdee. This type of inquiry has matured in diverse directions under the common name of string theory and its several manifestations.[3] The field has come under attack recently for not having yet produced tangible physical results[4] but there is little doubt that it has been a very stimulating construct that may ultimately begin to answer important physics questions.

Within this grand construct Salam placed himself in an important position — and rightly so. He had several abiding technical interests such as renormalizability, non-Abelian gauge theories and chirality. The importance of the Standard Model, which he helped shape, was realized more completely when Gerard t'Hooft proved its renormalizability in 1972 and the experimental confirmation came about in 1983 at CERN.

Physics has moved on. The recent major experimental developments in cosmology have introduced remarkable changes in the outlook of the Standard Model of quarks and lepton, and have deeply modified the views prevailing at the time of Abdus Salam. Even a normally conservative person today would say that we are witnessing a turning point. Recent experimental findings, which have led to the 2006 Nobel Prize to John Mather and George Smoot for their discovery of the black-body form and anisotropy of the cosmic microwave background radiation, have introduced an entirely new view of the constituents of the universe. It appears that the overwhelming majority of our scientific and technological knowledge has been confined so far to about 5% of the universe related to ordinary matter — both inanimate and living.[5] Determining the nature of the missing 95% of the Universe is amongst the most important problems in modern cosmology and particle physics — something that was unforeseen in Salam's time.

Changes in physics have come from another direction as well. Conviction is growing that reductionism, the cornerstone of much of 20th-century physics, has serious limitations of principle despite its enormous successes; that a deductive link does not exist between the finest constituents of matter and phenomena that occur on the human scale; that one needs an equally deep understanding of the so-called emergent phenomena regulated by higher organizing principles; that these organizing principles are equally deep in both content and structure. Perhaps it is too much to say that physics at the turn of the 21st century is undergoing a crisis similar to that at the turn of the last, but there is no doubt that the subject is changing its landscape.[6]

4. Salam's Concern for Science in Developing Countries

There is a second aspect of Salam's work that merits equal attention: his concern for scientists from poor countries — or developing countries as they are euphemistically called today. Towards the end of one of his lectures,[7] Salam remarked as follows:

> "Unquestionably, there has been no one like Einstein in physics of this century, but one has to reflect on how easily Einstein might have been lost, particularly if he had been born in a developing country ....
>
> Would an Einstein — with his total commitment to science for its own sake — fare well in the climate of today, even in a developed country, [in an environment that looks constantly for] social relevance, immediate applicability and cost-benefit analysis in supporting scientific research ...."

One of Salam's passions was that the best and the brightest in developing countries do not get lost because of lack of opportunities. Continuing from his description,[1] we have the following text:

> "I spoke earlier of the difficulties of doing science in developing countries. I would like to conclude with an appeal. Funds allotted for science in developing countries are small, and the scientific communities sub-critical. Developing countries must realize that the scientific men and women are a precious asset. They must be given opportunities, responsibilities for the scientific and technological developments in their countries. Quite often, the small numbers that exist are underutilized. The goal must be to increase their numbers because a world divided between the haves and have-nots in science and technology cannot endure in equilibrium. It is our duty to redress this inequity."

It was this passion that was instrumental in establishing ICTP as a center of learning where such opportunities might be provided for scientists from developing countries. Salam's specialization in high energy physics meant that the Centre was oriented initially towards that area of physics, but he never lost track of other branches of theoretical physics. The Centre had diversified in his own time as is evidenced by the following quote:[8]

> "The Centre has now existed for 18 years. It works on the interface of pure and applied physics and deals with subjects like the physics of energy, physics of materials, earth and environment, physics of microprocessors besides pure physics, the physics of energy, physics of fusion, physics of

reactors, physics of solar and non-conventional energy sources, biophysics, laser physics, microprocessor physics, communication physics, physics of the earth, oceans and deserts, and applicable mathematics, besides disciplines of pure physics like particle physics, astrophysics and relativity."

In keeping it with his own vision,[9] the Centre now does encompass several of these branches of physics and also mathematics. It is a lively place where ideas cutting across different branches of physical and mathematical sciences coexist, and has grown well past the confines of theoretical physics as it is generally understood. It is a hallmark of international cooperation in science working under a tripartite agreement among the United Nations Educational, Scientific and Cultural Organization (UNECO), the International Atomic Energy Agency (IAEA) and the Government of Italy (which funds a major share of ICTP).

Salam was well aware that physics is incomplete without an experimental component, and took interest in the experimental work of young scientists. In particular, one finds the following comments in a report that he prepared for the *ad hoc* committee evaluating ICTP in 1983:

> "There is a pressing request from experimental physicists coming to the Centre to find here at least some of the experimental facilities which are not available in their home countries. Two kinds of laboratories have been therefore proposed ... (a) Training and Demonstration Laboratories ... in which scientists could spend a training period ... and (b) Permanent Research Laboratories ... where high-level, modern research can be performed ...."

Thus, beginning around 1980, there has always been some experimental work at ICTP underlying Salam's belief that physics is the result of a fruitful interplay between experiment and theory (one has to include computer simulations these days). This has resulted in the creation of both types of labs mentioned above, and have included, at one time or another, microprocessors, aeronomy, distributed instrumentation networks, information and communication technology, optics and lasers, fluid dynamics, synchrotron radiation, high-T_c superconductors, materials science, accelerator physics, and so forth. Much of the experimental work has been done in cooperation with other local institutions in Trieste such as International Centre for Science and Technology, International Centre for Genetic Engineering and Biotechnology, the Synchrotron Laboratory Elettra, and Italian Institute for Nuclear Physics (INFN), and other institutions in Italy and elsewhere (such as the International Centre for Scientific Culture — World Laboratory in Geneva), as well as CERN. The best example of the interaction between theory and experiment is Salam's theoretical predictions and the experimental discovery at CERN, which led to the Nobel Prize for Carlo Rubbia and Simon van der Meer in 1984.

5. Salam's Broader Concerns

It is sometimes said that every great man has had at least one great idea. Salam may be said to have had two: the electroweak theory and the ICTP. As a physicist and as a human being concerned about poor countries and with scientists from there, Salam was simply admirable. He is one of my heroes, and I am honored to hold a professorship in his name.

As I said in the Preamble, Salam was a person with diverse ideas and drives. He was not content in trying out one single thing. I am therefore particularly unsympathetic to efforts that attempt to fit him into shapeless putty and forget the rich tapestry that made him the unique person that he was. In this spirit, I should point out at least one dimension of Salam to which I myself cannot relate — as indeed several others in his time could not. I have no special point to make here except to convey my bewilderment by the trouble it caused him. This dimension concerns his pronouncements on a wide range of subjects, such as the history of science across cultures and ages, and the grating admonitions of the rich and powerful. There was often more rhetoric than substance in these deliveries, and generalizations more sweeping than to which he was entitled on the strength of cursory sources that seemed to have been consulted.

It is even more difficult to appreciate his latter-day preoccupation with Islam, his penchant to proclaim religiosity, and the drive to proclaim that he was a believer and a practicing Muslim — sometimes attempting to establish that he was better at it than others. These extraordinary circumstances, probably in part the result of the religious persecution that he indirectly and directly faced, did not prevent him from being excommunicated eventually: I have in possession a letter in which he remarks on this fate with extraordinary sadness. It was also clear that he met insurmountable hurdles from the officialdom of his country when he made concerted efforts to become the Director-General of UNESCO. That his health deteriorated soon after this failed attempt is perhaps no coincidence, though it is hard to prove the connection. That no one in the Pakistani power structure at the time felt free to attend his burial, and that his remains lie buried in a grave of no consequence, are sad facts that one cannot but reflect upon glumly. I have been told that Salam was never allowed to make his Hajj[10] and that, in an incongruous and meaningless attempt to exclude him from Islam even in his death, the words on his grave "there lies the first Muslim Nobel laureate" have apparently been altered[11] to "there lies the first Nobel laureate".

Somewhere in the kind of end that befell a great and passionate man lurks a lesson. Perhaps religion and science cannot be mixed indiscriminately, though one can live them simultaneously and successfully in one's life without making a hard sell of either. If Salam's message was that science does not negate spiritual outlook, it is indeed a valuable point to drive home — especially in our era in which considerable concern exists that scientific outlook is somehow a negation of spirituality. If his thesis was that religion is not blind faith, it underscores an eternal view. But the coexistence of science and religion can be imbued only through example that sanctions no aberrant proclamations. One of the most profound statements I know is that it is hard to tell apart, through causal encounters, a deeply spiritual person from one who is not.

6. Final Remarks

One of Salam's well known quotes, adopted as one of ICTP's driving mottos, is that "scientific thought is the common heritage of mankind". In the scientific legacy of our species, many countries and cultures have indeed made crucial contributions — some, no doubt, more than the others. This subject is worthy of deep study and cannot be reduced to clichés. Salam's core concern was that science had become the province of the West in recent three or so centuries, and that the situation needed to be brought back to normalcy (that is, a situation in which all cultures contributed to science) if the world as a whole were to share its benefits. He particularly bemoaned the fact that science in Islamic countries had fallen on hard times, and, both privately and publicly, cajoled Muslim scientists to change the situation in all possible forms. Unfortunately, his considerations on this score remain valid, by and large, even today.[12]

One should, however, not forget the reasons why the West has been able to gain the ascendancy in science and technology. The West is not innocent in how it has appropriated a good part of the world's wealth and resources; indeed, there is no doubt that this propensity has played a major role in its recent rise to power and plenty. Insofar as it concerns science, however, this pre-eminence lies in its ready acceptance of factual evidence, wherever it may have come from and wherever it may lead to; the courage to make risky hypotheses but the willingness and discipline to subject them to the rigor of experimental verification; a strong focus that does not permit solace to be found in subjective experiences or in the authority of a text. It is not as if the West of yesteryears, or of today for that matter, is flawless in its pursuit of truth — one only has to recall the fate that befell Galileo and the modern-day rise of creationism. Even so, the underlying qualities remain as stated in so far as it concerns the best science that we have inherited. It is the willingness — indeed eagerness — to challenge and be challenged that allows us humans to build the basis for comprehending the universe and our place, if of any consequence, in it.

If the rest of the world catches up on these traits, Salam's dream in its best sense will have come true. The institution that he created, namely the ICTP, and those of us who have followed his footsteps and tried to fill his large shoes, will be proud to be part of his dream.[13]

References and Endnotes

1. From "One hundred reasons to be a scientist," published by the Abdus Salam International Centre for Theoretical Physics, Trieste, Italy. A copy of this book for a nominal cost, as well as items 10 and 11, can be obtained by writing to ICTP library at library@ictp.it.
2. See, e.g., *Selected Papers of Abdus Salam*, edited by A. Ali, S. Isham, T. Kibble and Riazuddin (World Scientific, 1994).
3. There are now several popular books on string theory, and I cite just one example. B. R. Greene, *The Elegant Universe: Superstrings, Hidden Dimensions, and the Quest for the Ultimate Theory* (W. W. Norton, New York, 1999).
4. See, e.g., P. Woit, *Not Even Wrong: The Failure of String Theory and the Search for Unity in Physical Law* (Basic Books, New York, 2006).
5. I am not aware of a comprehensive account from which to gain a full perspective of the subject. Popular accounts can be found in science magazines such as *Scientific American*. The article by R. R Caldwell, "Dark energy," published in *Physics World*, in the May 2004 issue, is one such.
6. Even though the idea of emergence is no longer new, it is frustrating that one cannot recommend a concise source for gaining a decent understanding of the subject without succumbing to hyperbole and mushiness. Perhaps the article by P. W. Anderson, "More is different: Broken symmetry and the nature of the hierarchical structure of science", published in *Science*, 177, 393–396 (1972), may still be regarded as a good account of the main points.
7. Speech given at UNESCO by A. Salam on 9 May 1979, commemorating the 100th Anniversary of Albert Einstein.
8. Address given by A. Salam in Belgrade, 13–17 December 1982, in a meeting on the establishment of ICGEB.
9. Salam's vision was actually different in detail. He had envisioned establishing separate institutes on each of these aspects of science, at least some of them in developing countries, all of them federated in some fashion with ICTP. In 1969, in Nobel Symposium 14, Stockholm, 15–20 September 1969, he delivered a talk under the title "The advancement of science for the developing countries". In part, he said there as follows: "... Such Institutes, together with the International Institutes I have spoken about in the developing countries — both in pure and applied sciences — as well as in Economics, Sociology and other studies of man — would make up a Federation, enriching each other by contacts, deriving strength from common ideals shared and practised. As I said I would like to see such a Federation linked up with the United Nations Organization or one of its Agencies in a loose connection. ... Before this Federation of World Institutes begins to look like the World University ... new Institutes ... will have to be created to link up with this Federation. But the first step — the Federation — could perhaps come even within the next ten years." It is not exactly clear when he came to realize that this grand vision was beyond realization. However, it seems to be the result of this realization that ICTP since Salam's time began to expand its coverage of areas other than theoretical physics, interpreted narrowly. Indeed, it seems to be the direction to go.
10. F. Hussain, "Salam, Saudi Arabia and Pakistan," can be found at the website http://bznotes.wordpress.com/2006/06/23/salam-saudi-arabia-and-pakistan-%E2%80%93-a-disgrace-by-faheem-hussain/.
11. N. Subramanian, "The scientist that Pakistan chose to forget," *The Hindu*, November 31, 2006, can be found at the website http://www.hinduonnet.com/2006/11/30/stories/2006113004621000.htm

12. See the following two articles that appeared more or less simultaneously. P. A. Hoodbhoy, "Science and the Islamic World — The Quest for Rapprochement," *Physics Today*, August 2007, pp. 49–55; and a profile on R. Mansouri, "A way forward for Islamic science," *Physics World*, August 2007, pp. 12–13.
13. Dreams have to go together with efforts on detail. I shall make this remark with no prejudice. In the part of the world from where Salam and I come, dreamers have been plenty but the ones that make the dreams come true are few. There are many reasons for this mismatch but let me quote from William Blake, regretfully without reference, to make at least one point: "He who would do good to another must do it in Minute Particulars: general Good is the plea of the scoundrel, hypocrite, and flatterer, for Art and Science cannot exist but in minutely organized Particulars and not in generalizing Demonstrations of the Rational Power." I am not a great fan of Blake but this particular statement rings true.

Salam the Man

SALAM'S MANTLE OF BRITISHNESS

GORDON FRASER*

CERN, Geneva, Switzerland

As the final speaker today, I am limited because others have already covered most of Salam's major achievements. So after having marked the 50th anniversary of Salam's arrival at Imperial College, I will go back another ten years or so to 2 October 1946, when a motivated, but cold and confused student arrived at Liverpool docks, wondering what he should do next. As it happened, he had no need to worry.

Salam's scientific career was built on a British foundation. Tom Kibble has already covered the key role of Paul Matthews, and I will point out some of the other people and institutions to help and influence Salam. In October 1946, he was en route to St John's College, Cambridge, to continue his mathematics studies, after already having studied mathematics for four years at Government College, Lahore.

To get to Cambridge had been a major challenge. During the Second World War, the government of his Punjab district had dutifully collected money to support the war effort, but at war's end, some of this still lay unspent. The idea was to use this cash for scholarships to enable sons of poor farmers to study overseas.

But Salam's father was a school inspector, not a farmer. To qualify as such, the first requirement was to have land. His brother (Salam's uncle and future father-in-law) immediately donated a tiny plot, large enough to keep a cow. With additional land and animals on it, Salam's father technically became a farmer, so Salam qualified for a scholarship.

Meanwhile he had already set in motion the mechanism of the British university admission process, and on 3 September 1946 received a telegram saying that he had a place at St John's College. He knew that he had to move fast, but was unsure exactly which direction to go.

*Gordon Fraser was a research student in Salam's group at Imperial College London. After a career as a science editor and writer, mainly at CERN, he is now a freelance science author. His biography of Salam *Cosmic Anger, Abdus Salam — The First Muslim Nobel Scientist* was published by Oxford University Press in April 2008.

First he had to confirm that the scholarship money was available. Salam took the overnight train to Lahore to visit the Punjab Education Department, only to learn on arrival that the department decamped for the hot summer to the hills of Simla, 250 kilometres away. When an anxious Salam finally arrived at the Education Department, he was overjoyed to meet a messenger with a letter for him. Salam indeed had money to go to Britain. The offer was conditional on having a university place, but Salam now had the telegram from St. John's in his pocket.

Salam was supposed to begin his studies in Cambridge in October. But first he had to get there. The boat trip would take several weeks. With no time to lose, he took the train from Simla to Delhi, another 250 kilometres. By the time he arrived, the offices of the shipping company had closed for the weekend. After an argument, a reluctant clerk gave Salam a provisional booking. He then took the train back to Multan and spent the next day packing.

Salam said goodbye to his family, and departed for Bombay, 1000 kilometres away, another two nights on the train. There he would wait for a boat. Salam's father stayed at home to pray. On arrival in Bombay, the city was under curfew and Salam bolted into a run-down hotel. No sooner had he fallen asleep when he was woken by pounding on the door. The British Military Police were looking for deserters from the Indian navy. After producing his Cambridge University letters, Salam was able to go back to sleep. The next morning he fought through the crowds at Bombay Docks and was assigned a berth.

When his boat arrived at Liverpool, the disoriented Salam recognized on the dockside an impressive figure. Sir Muhammad Zafrullah Khan was a lawyer and the most prominent member of the Ahmadi Muslim community in British India. He had not come to Liverpool to meet Salam: he was in the UK en route from the United States, and had paused in London before returning to India. He had come to Liverpool to meet a relative, arriving from Bombay on the same boat. In the aftermath of the Second World War, the United States was becoming the dominant Western Imperial power, taking up the slack left by war-impoverished Britain, which was preparing to quit India. The United States wanted to know how this would happen. Zafrullah Khan had met President Truman in Chicago.

Later, after the partition of British India, Zafrullah Khan was to become Pakistan's first Foreign Minister, and later its permanent representative at the United Nations. Because of anti-Ahmadi prejudice, he left Pakistan to join the International Court of Justice in the Hague, where he was to serve as its President. Later in Salam's career, Zafrullah Khan introduced him to the world of international politics in general and the UN in particular. Without Zafrullah's prior guidance, Salam could never have achieved his objective of setting up an international science centre under the banner of the United Nations.

At Liverpool docks in 1946, Salam had 25 kilos of mathematics books, but was otherwise unprepared for undergraduate life at Cambridge. Seeing his predicament, Zafrullah took off his heavy overcoat and gave it to the young student. This gallant act gives the title for this talk.

In Lahore, Salam had been an avid reader of books, but no amount of reading could have prepared him for post-war Britain, victorious but debilitated after seven years of conflict. It was also a nation in transition: the new socialist government was shedding an imperial past as fast as it could, and nationalizing great swathes of industry. Soldiers were returning to families which hardly knew them, and there was a nationwide housing shortage. In addition, nobody knew that global warming had yet to set in, and stricken with fuel shortages the country had to experience its coldest winter for 140 years.

In the midst of all this change, undergraduate life at Cambridge provided a relatively safe haven for Salam. During his five years at Cambridge, in some kind of curious imperial hangover, Britain was to win the Nobel Prize for Physics four times. This surely impressed many people, including Salam, who arrived at Cambridge as a mathematics student, but would leave as a research physicist.

Among those late 1940s Nobel Physics laureates was Patrick Blackett, who brought Salam to Imperial College, the reason why we are here today. Blackett epitomized the fortunes of 20th-century Britain, moving from an initial career as a naval officer to becoming the first person to witness a nuclear disintegration. After the Second World War, where he had done important work in operational research, he became scientific advisor to Indian Prime Minister Nehru. Having done his duty to one of the new nations of the subcontinent, perhaps he felt that he should also help the other. Arriving at Imperial College to spearhead its postwar expansion, Blackett set about headhunting. Among those he found were Dennis Gabor as Professor of Electron Physics (Gabor was to win the 1971 Nobel Physics Prize for his development of holography) and Salam. Later, after establishing his new centre at Trieste, Salam had difficulties when negotiating his ongoing status at Imperial College. Blackett supported Salam's request for secondment: "the [Trieste] Centre is Salam's creation and it might not flourish without him", he urged.

Salam's appointment as Professor of Theoretical Physics at Imperial College was the second time in a few years that he had been offered a British job for which he had not applied. The first had been as Stokes Lecturer in Mathematics at Cambridge, and the person responsible for that was Nicolas Kemmer. Like Salam, Kemmer had also assumed a mantle of Britishness after having been a victim of prejudice and intolerance. In Salam's case, it was because of his status as an Ahmadi Muslim in Pakistan. Kemmer had had to leave Nazi Germany, and took his first steps in refugee research in the UK at Imperial College. Later he became Stokes Lecturer in Mathematics at Cambridge. There, Dirac traditionally shunned research students, so Kemmer had to deal with the backlog and sometimes felt overburdened.

When Paul Matthews was about to leave Cambridge and embark on an independent research career, Kemmer was pleased to have shed one research student. Then Salam, who had initially been earmarked for experimental research because of his remarkable exam performance, came knocking on his door asking to be taken on as a research student in theoretical physics. The overloaded Kemmer initially cold-shouldered him, but later, after seeing Salam's success with renormalization theory in partnership with Matthews, felt guilty. Thus when Kemmer left Cambridge to take up Max Born's chair in Edinburgh, he advised the powers-that-be to offer Salam, by then back in Lahore, the job of Stokes Lecturer.

But back to Salam's undergraduate career. He had already done undergraduate mathematics and a master's degree in Lahore. Why was he redoing relatively elementary undergraduate mathematics at Cambridge? Salam's father had long told his son to aim for the Indian Civil Service (ICS), which was staffed predominantly by talented young Britons who managed commendably to administer the affairs of the sub-continent. In principle the ICS was open to Indian candidates, but selection was difficult, and a helpful pre-requisite was to have been educated in Britain. During the Second World War, ICS recruitment had been put on hold, and in any case would be stopped by British withdrawal from India and its partition into two independent nations. But whatever the outcome, the new nation of Pakistan would still need skilled administrators, and this was the career path that Salam's father foresaw for his son.

Salam's undergraduate mathematics tutor at Cambridge was the mercurial Fred Hoyle, a sort of British equivalent to Richard Feynman — bright, not embarrassed by a humble background and even abrasive, unafraid to speak his mind. In the late 1930s, Hoyle had become one of the few to achieve the status of Dirac's research student, mainly because Hoyle, whose research was going well, had to find a replacement formal research supervisor when Peierls, his first supervisor, left Cambridge. Both Hoyle and Dirac could not resist the irony of a student who did not want a supervisor and a supervisor who did not want a student.

After two years of undergraduate mathematics and a first-class degree (a "wrangler"), Salam was faced with a dilemma — should he continue with Part III mathematics or switch to physics? Salam was now highly motivated to opt for the latter, but still felt the responsibility of the civil service goal stressed by his father. Confused, he asked Hoyle, who advised "do what you think would be best for your country", adding "if you want to become a physicist, even a theoretical one, you must do the experimental course at the Cavendish Laboratory. Otherwise you will never be able to look a theoretical physicist in the eye".[1]

G. P. Thompson (the son of J. J. Thompson) and Nevill Mott had tried for a first-class Cambridge degree in Physics Part II in one year after completing Mathematics Part II. Both failed, but G. P. Thompson shared the Nobel Prize in 1937, and Nevill Mott would emulate the feat exactly 40 years later. Achieving a first-class Cambridge physics degree in just one year after having first done mathematics looked harder than getting a Nobel Prize. Salam's name went down for Physics Part II, to see if he could do better than Thompson and Mott. He did. The Nobel Prize would take care of itself later.

Why did Salam want to switch to physics in 1948? The news of Blackett's Nobel Physics Prize, one year after Appleton's, must have been an influence. But in doing Mathematics Part II in his second undergraduate year, Salam had followed the lectures of Paul Dirac. Salam said later "Dirac represented the highest reaches of personal integrity of anyone I have ever met". Although Dirac took few research students, his major impact was via his course of lectures on quantum theory. Undergraduates, novice graduate students and professional researchers would all attend, admiring the remorseless flow of the master's logic. In his obituary of Salam, Hoyle later wrote "For Salam, the greatest scientist of the 20th century was undoubtedly Dirac. Of course, you could say this was one St. John's College man supporting another. But when I asked [Salam] if this included Einstein, he was clear in his answer: 'Einstein had his mathematics all done for him. Dirac invented his. Not only that, but it was Dirac who first made it clear that the route towards real understanding in theoretical physics lies through abstract mathematics.'"[2]

Later, Salam said that if he had known that his future lay in research, he would have tried to go directly from Lahore to a research degree.[3] One who did so was a contemporary of Salam, a student at Punjab University, Har Gobind Khorana. Born a Hindu in the tiny Punjabi village of Raipur, Khorana progressed to Multan High School and then Punjab University in Lahore, achieving a chemistry BSc in 1943 and an MSc in 1945. With a Government of India Fellowship he came to Liverpool, going directly into research, gaining his PhD in 1948. In 1950–52 at Cambridge, he briefly, and again unknowingly, overlapped with Salam. Khorana went on to share the 1968 Nobel Prize for Physiology and Medicine (with Robert W. Holley and Marshall Warren Nirenberg) for their work on the interpretation of the genetic code and its function in protein synthesis. Khorana and Salam never met.

After his Nobel Prize, Salam's mantle of Britishness was complete, and he became the elder statesman of British subnuclear physics, despite the fact that he spent most of his time in his centre in Trieste. The UK, having shed its imperial past in the immediate postwar years, was looking to shed more. In the 1970s, the UK streamlined its national research effort by discontinuing the home-based national particle physics research programme based on synchrotrons, outsourcing this research to the pan-European effort at CERN in Geneva. After the glorious tradition of Rutherford earlier in the century, it had been the scientific equivalent of quitting India.

With a once-proud home-based research programme gone, in 1984 the British Advisory Board of the Research Councils commissioned a review group. In those cost-conscious times, money for CERN and other international Big Science ventures had become highly visible, and other sciences were clamouring for a similar share of research funding. Salam was influential in maintaining British membership of CERN. "Particle Physics: Will Britain Kill Its Own Creation," he thundered in the *New Scientist* on 3 January 1985, concluding "Clearly, a country that has upheld fundamental science ... cannot lightly absolve itself and withdraw from supporting this most exciting adventure of ideas of our times."

Salam took his British responsibilities very seriously. In 1981 he was apparently the only person in the UK to notice that it was the 150th anniversary of James Clerk Maxwell, the pioneer of electromagnetic theory. Unfortunately even Salam left it too late, and the anniversary went almost unnoticed. The pinnacle of Salam's recognition in the UK came in 1989, when he was made an Honorary Knight Commander of the Order of the British Empire (KBE). He was the first non-British scientist to receive the award.

There are interesting parallels between Salam and the other 20th-century mathematical giant of the Indian sub-continent — Srinivasa Ramanujan. Both came from modest backgrounds, and both achieved intellectual fame by coming to Britain. Ramanujan's aspirations were limited to mathematics, but mathematical science and Britain were only two corners of the world stage on which Salam operated. The platform of Salam's endeavours rested on several pillars: to further our understanding of the underlying workings of Nature; to create an infrastructure for Third World scientists; and to advance the cause of science in Islam and restore it to the glory of a thousand years ago.

Salam indeed furthered our understanding of Nature, as his Nobel Prize attests. But if he had not been, others would have still achieved what he did. However nobody else built an infrastructure for Third World scientists: this is his true monument. And as for advancing the cause of science in Islam, his continual efforts were undermined by the fragility of his Ahmadi perch. But the sum total of these achievements adds up to far more than those of Ramanujan. While the mathematics genius from Madras has become a legend which motivates generations of students and inspires books, plays and films, Salam is little remembered away from the Trieste centre which bears his name. The story of a young lad from a market town in the Punjab who became a leader of science and emerged as a champion of the world's poor deserves at least as much recognition as Ramanujan.

I would like to thank Professor Michael Duff for inviting me to speak at the Salam 50th anniversary event and for his kind hospitality at Imperial College.

References

1. A. Salam, *Science Sublime*, in *A Passion for Science*, eds. L. Wolpert and A. Richards (Oxford University Press, Oxford, 1988).
2. Sir Fred Hoyle, Obituary to Abdus Salam, *The Eagle* (St. John's College, Cambridge, 1997), pp. 80–85.
3. A. Salam, A Life of Physics, in *Frontiers of Physics, High Technology and Mathematics*, eds. H. A. Cerderia and S. O. Lundqvist (World Scientific, Singapore, 1990).

RECOLLECTIONS OF MY FATHER

AHMAD SALAM
(Son of Abdus Salam)

In the Name of Allah, Most Gracious, Ever Merciful

Sir Richard Sykes, Professors Mike Duff, Tom Kibble, Kelly Stelle, Arttu Rajantie and Chris Isham, Her Excellency the High Commissioner, friends, family, colleagues, ladies and gentlemen,

It is indeed an honour to be able to stand before you today here at Imperial College and say a few words about my late father. Thankfully I only have a few minutes ... so that makes my task a little easier and yet also a lot more difficult. What can I tell you about such an incredible man that you don't know already? You all know about his science, his humanitarian work, his deep religious beliefs and motivation. You all know about his simplicity and humility and his total and utter dedication to the cause of equality and opportunity. So what is left that I can tell you?

When Mike Duff asked if I would say a few words at today's meeting, I of course had to say yes. But then I thought it may be more interesting to make this more like a TV programme, in this case "An Audience with Abdus Salam's eldest son". I thought this could make it more interesting for you the participants, and all I would need to do was stand here and answer any questions you may have.

But I didn't think I would get off that lightly and I certainly didn't think this was in the spirit of Abdus Salam. It may have been an easier option for me and I am sure it would have been more entertaining for you the participants.

But then I thought long and hard and tried to remember any time that father had taken the easier option? Apart from when he tried to make coffee, lunch, tea or tried to cook in general, or do any kind of domestic chores ... which thankfully was not too often ... he didn't know the easier option. His whole life revolved around constant hard work, effort, struggle and striving for the maximum. He never let up. He never ever gave up.

He didn't take holidays: he didn't know what they were for;

He didn't take a day off in the week; he simply didn't stop.

He would always take his work with him, planes, trains, cars, lunches, dinners, lectures, hairdressers. There wasn't anywhere he would not work or scribble down some ideas or theories. His mind was always working. There is of course the wonderful story about lunch at Buckingham Palace with the Queen and Prince Philip. After the lunch when her Majesty, the Corgis and all the other guests had left, he popped back into the room to ask if he could have the napkin as he had written some notes on it.

Imam BA Rafiq of the London Mosque tells the story of how when father was at the Mosque for Friday prayers, he was seated in the front row and after a while Imam Rafiq saw him take out a notebook and begin to write in it. Imam Rafiq was of course very pleased as he thought father was taking some notes of the sermon. So after the service was over he went to ask my father what he had liked about the sermon and what he had written down? Father replied he had had a physics idea and wanted to record it there and then.

Father was a very thorough and methodical man with a great routine and set likes and dislikes. He would always research extensively, prepare, read, study, research more and then finally be ready to say what he had to. He was a firm believer in doing what he felt was right, where others agreed or a disagreed. He would listen to others and then usually go with his own opinion. This single-minded (some would call it stubborn) approach was I think something he inherited from his father.

My grandfather was the greatest single and most dominant influence in my father's life. His total and utter devotion to his parents and their welfare was a wonderful example to us all. I am not sure that many of you know that the Nobel Prize money was put into a charity in memory of my father's parents for the education of young children in developing countries, irrespective of colour, creed or nationality. That fund still exists in a small way today.

As some of you may be aware my grandfather had been foretold in a dream that he would have a son, and he was told to name him Abdus Salam. This was a great sign of impending greatness, and when in another dream grandfather saw a young boy climbing a very tall tree that disappeared high up into the sky. When my grandfather called on this child to stop and come down, the child looked down and smilingly said "don't worry father" and continued so high that he became invisible. This was another sign that father would achieve great and lofty successes and thus my grandparents made father the primary object of their attention. Father was the blessed son, the one whom grandfather and grandmother doted on and spoilt so that he was given priority in everything. He was excused from household chores such as milking the cow that his brothers had to undertake, he was given the best meals that were available, though to be honest in such a poor family that was not much. Grandfather would sit and teach and read with young Abdus Salam and encourage him to always be ahead of the class. This was a lesson that I too was taught at a young age.

Father was always very keen to teach me maths and physics as much as he could himself; but he wasn't a very good teacher of 7-year olds. I remember one of his students later in life telling me that he would always attend every one of my father's lectures at Imperial, not because he understood a word of them, but simply because it inspired him to go and learn about the subject more.

But I digress; father had a certain notorious impatience. If I did not get the workings of a mathematical problem first time he wouldn't have the patience to talk me through it. After a while he recognized that teaching me was a pointless task and waste of his very precious time and it was easier to let someone else teach me; and I have recollections of having a few tutors who would teach me usually after school on a Saturday. I recall one in particular, an Indian student here at Imperial. I think he said yes because he didn't dare say no to father. Actually very few people said no to father ....

So every Saturday, after school finished I would get the bus from Richmond, and come to Queens Gate for my lessons here at the College. So I can claim I studied Maths at Imperial!! By the way this was not unique to Imperial and London, when we went to Trieste for our holidays, father would persuade a poor suffering individual to teach me whilst I was there on a so-called holiday!!

Yet my first introduction to Imperial begun in the late sixties when as a young boy I would come with father to the College and sit time with him in his room playing whilst he worked behind a huge desk loaded with papers. Because I saw so little of him it was a special time for us to be together. He was very, very proud of his position, his role and time at the College. Imperial held a very special place in his heart.

I was allowed to draw on the board providing that I didn't rub anything out ... it was a long old-fashioned blackboard which covered one full wall and it was great fun to scribble in chalk for hours!

I was always amazed at the amount of files and papers on father's desk, but he always seemed to know where each document and piece of paper was. This was a quality which never deserted him; I recall throughout his life he always knew where something was; a book, a pen and notepad or whatever. He was a very organized and ordered man. There are many stories both in London and Trieste where he would ask somebody to get a book on such and such subject. He would say it was a grey book on the second shelf on the left on, for example Russian ballroom dancing ... and invariably he was correct. It was on the second shelf on the left. When we packed for him for one of his many journeys, everything had to be out into a bag. So notes, papers, clothes, shaving gear, etc. all had their own bags.

He would tell me that organizational ability is an asset in life and for great enterprise it was a necessity.

Speaking about books, I think we all know he had a voracious appetite for books. He did actually buy a book on Russian ballroom dancing and I have to say to my eternal regret I never had the courage to ask him why he bought it. I think we have kept that book amongst a few other unusual choices such as Teach Yourself Air Navigation and the Memoirs of Groucho Marx! Father loved reading and was an avid speed reader. History and biographies greatly interested him; current affairs were also of great interest. He was also a major customer of secondhand bookshops and for many he alone provided up to 90% of their income. I think for a few years he was the best customer of Karnac booksellers on Gloucester Road here in Kensington. He would spend a great deal of time in these bookshops and I recall even when he was confined to a wheelchair, that I would take him through the bookshop at Heathrow so he could choose books and magazines which he would buy with an obligatory bar or box of chocolates. As children we were always encouraged to read extensively and were always allowed and encouraged to try as many books as we wanted, a lesson my children benefit from today.

As I said father was very proud of Imperial. It had almost given him his first recognition, post Cambridge and after his heart-breaking experience in Pakistan, Imperial provided him with much needed support. The undying support of the staff and authorities at Imperial was a major event. Just imagine how many colleges today would give permission for one of their Professors/Head of Dept to be outside the country for between 15 to 20 days a month? It is a tribute to the farsightedness of the College authorities that they fully appreciated it was worth having even a small percentage of something very special rather than nothing at all. And I know father was very aware of this very special relationship and he took great pride in flying the college colours and promoting Imperial whenever he could.

Father had a real eye for a bargain, especially if it was in the newspapers. We would often come down to find the morning paper in shreds as he had torn out an advert for a 6 waveband 300 station Russian Radio, ex Army surplus for someone knock down price. Perhaps he was trying to find the station that played Russian Ballroom Music! I have bought one such radio with me on display on the table. More often than not father would tune into the BBC World Service followed by Radio Pakistan to pick up any news he could on his beloved country. He never ever turned his back on Pakistan, no matter what the politicians did to him, he remained 100% loyal. Despite many offers of nationality he never ever gave up his Pakistani nationality and was deeply proud of being a Pakistani. I remember after he had won the Nobel, he designed a tie with the logo BBP (Build a Better Pakistan). This love was reciprocated by the people. I will never forget the literally thousands who came for his funeral prayers; people had walked, come by bus, rail, car, truck anyway they could to pay their respect to their local boy who made good.

Father also had a wonderful eye for a bargain; if something was cheap enough then he would buy more than one, so I think we had four of the Russian radios. Or in the case of bargain Chinese made shirts it was literally two dozen. My wife says I have inherited this habit! He had very simple and ordered tastes in everything; food, clothes, etc. When in London, every night my Mother would place a tray of tea in a thermos flask, Digestive biscuits and Ritz crackers, and some cheddar cheese, a bowl of cashew nuts and some bananas for father to have an early snack at around 3.30am when he woke up. He would rise, pray and then start his work. For him this was the perfect time to work completely undisturbed. No phone calls, no people and no noise. He would have the Holy Quran playing in the background; his room was always very warm with a strong smell of incense which he burned incessantly. He would finally rise at about 8am and have a full breakfast; beef kosher sausages, (halal was virtually impossible to find in the sixties) or fried eggs on toast, later in life he moved onto fish fingers and then he had a short-lived interest in muesli because he had read somewhere the benefits of muesli. This was washed down with copious amounts of hot sweet tea. He had developed a taste for a very nasty drink called Camp Coffee; it was not coffee as we know in any shape or form but chicory essence! Nasty and very bitter. He kept trying to convince guests who came to visit him and us that it was actually palatable ... trust me it was awful. I have kept a bottle as a souvenir.

He had a well developed sense of humour, derived largely from PG Wodehouse, possibly his favourite casual reading author. Every room in the house, including the smallest, would have a PG book somewhere. He kept trying to persuade from an early age that I would enjoy it and find it very funny. Sadly for him I don't think any of us children enjoyed PG as much as father. I know I certainly didn't. One of my fondest memories was on the very rare and odd occasion that father would watch TV, he would sit in his favourite armchair and watch an old BBC programme called Dads Army, this was a light comedy about a local defence platoon set in World War 2 in a corner of rural England led by a very pompous captain. The programme would literally make father laugh and fall off the chair. We loved watching it with him; though for us the pleasure was to see him in such a relaxed and happy frame of mind. He also loved the early Charlie Chaplin films, the Buster Keaton, Laurel and Hardy films and all the Marx brothers.

I could go on and tell you many many stories and recollections, but I know we have a full programme and I am sure you would like to hear the other speakers.

I wanted to end with a very special tribute. We all know that behind every great man is an even greater woman; in this case my mother. Sadly my mother passed away just over three months ago and she had been looking forward to this event. I don't think my father would have been able to achieve as much as he did without her. From the very first moment when her father gave his brother, my paternal grandfather the small piece of land which made my father eligible for the scholarship which brought him to Cambridge, to the very end when she cared and looked after him during his tragic illness; she had supported him throughout all the good and bad times. She let him go and do all he wanted and needed to and took care of everything back at home, raising 4 children single-handedly, supporting them and him and providing for them and him in every possible way.

I think if father was here today he would recognize her contribution and pay her a special tribute.

Finally, may I extend a vote of thanks on behalf of the Salam family, all the members of the audience here today and to those who could not be here, to the hard work of the organizing committee. I know how much work Mike Duff, Tom Kibble, Chris Isham, Kelly Stelle and Arttu Rajantie have put into arranging and hosting such a wonderful commemorative day.

Father would have been honoured, delighted and humbled by your gesture. On his behalf I thank you all and of course the unsung heroes in the back rooms, the secretaries, administrators and others who made this possible.

Thank you very much.
May Allah bless you all.

After his death I was going through some of his own personal writings and thoughts; things he thought may never see the light of day ... some of the quotes illustrate his state of mind:

> "I am claiming that I am working. Tis false. Life is as soft as it ever was. Make it harder."

On the same page was another:

> "In the present there can be no reason for fatigue. It is quite true. Having worked constantly for many days, there is no longer any feeling of tiredness."

Another:

> "My two biggest enemies; light reading, light pursuits at hand and the feeling of light tiredness and light emptiness."

He was always striving to excel and to challenge and to get his message over. Another quote I found was

> "To write, to learn to write, to convey to others the vision of your knowledge, your learning, your wisdom. To write fluently, effortlessly."

He was a great communicator and a great leveller, be it literally a King, or a bathroom attendant. Another of his quotes which illustrates this is:

> "I have just begun to appreciate the integrity, the painstakingness of the smallest newsvendor to that of a University Dean, the vision of values, the deep moral sense, the lack of cynicism which ever makes for a good living."

RECOLLECTIONS OF MY FATHER

AZIZA RAHMAN
(Daughter of Abdus Salam)

Honored guests, ladies and gentlemen,

What do you say when people you hardly know come up to you and ask "What was your father really like? Or better still, "What's it like being the daughter of such a great man?" I never knew how to answer those questions, and I still don't. All I can do is to share a few of my favorite recollections with you and to simply say that I was very blessed to be the daughter of such incredible parents.

I am sure all of you know that my father's extraordinary story began when my grandfather, Chaudry Muhammad Hussain, was given the news of his birth and his name Abdus Salam, in a vision on June 2, 1925, while offering his evening prayer.

Consequently my father was born on January 29, 1926, in Jhang, which was at that time a small insignificant village in the Punjab, without electricity or any other amenities. His family was not rich in possessions, but was very rich in traditions of piety, scholarship and religious learning.

My father actually began his illustrious career by winning his first award at the age of two; for being the "healthiest" or rather the "fattest" baby in Jhang! Of course, there was no stopping him after that. I am honored to say that I don't think there was any distinction, award or honor that was not conferred on him right up to the end of his life.

My earliest memories of my father date back to when my mother and I moved to Cambridge with him. I was three or four years old at the time. We lived in a small flat near St. John's College, where my father was working. He had chosen to work at St. John's, turning down an offer from the more prestigious Trinity College, because the gardens of St. John's were prettiest in Cambridge. One of his favorite pastimes was to take us punting on the river. My mother and I would sit in the front of the punt, while my father would manage the pole; and I can remember more than one occasion when the pole would fall into the river and we would end up paddling the punt to the shore.

A few years later, when my father was appointed Professor here, we moved to London and Imperial College became a part of our lives too. Soon it became a Sunday morning tradition to accompany him to the college for a few hours while he worked. I remember walking down the long corridors of the old Physics building to his office, where we would be allowed to draw on the blackboard and do pretty much anything, as long as we were quiet. Later on, when my brother Ahmad was old enough to accompany us, my father would drop us off at the Science Museum, where we would spend the time playing with those old railway train models which would move up and down the tracks in their glass cases at the push of a button. Sometimes it would be the Natural History museum, where we would gaze in disbelief at the size of the blue whale. And the best part was on the way home, when we made that essential stop at Oppenheimers, the bookstore by South Kensington Station; where my father would load up on all kinds of books, and I could pick as many as I wanted. And sometimes, if he was in an particularly good mood, there would be the special treat of going to Wimpy's across the street.

In spite of his busy workload and increasingly frequent trips abroad, my father always made time to supervise our education. He spent a lot of time preparing me for the 11 plus exam, and was very pleased when I managed to do well and got a scholarship to Putney High school. He always took a keen interest in our education and our schools. Soon after I went to Putney High, the school held a fundraiser. Having just read a book on palmistry, my father volunteered his services as a palm reader and dressed in his sherwani and turban, was the hit of the fete, I wonder how many people who had their palms read that day would have believed that the palm reader was a future Nobel Laureate!

Our school holidays were always filled with schedules of study and workbooks to be done. My father set very high standards and whenever he returned from a trip, we would be called into his room for the dreaded check on our grades and progress.

Every activity had to have some educational value to it. Once, to our surprise and delight, he decided to take us to see the film "Lawrence of Arabia". It was the first time we had ever been to a movie theater. However, at the intermission he decided that enough time had been spent on it and that we should go home. Terribly disappointed, we begged him to let us see the rest. He agreed, but he himself left and waited in the car for us, where we found him deeply engrossed in work. When we got home, we had to write essays about TE Lawrence to show what we had learnt.

Sometimes he would decide to take a break from work and take us to Richmond Park just for a drive or a walk. Never one for completely switching off, my Father would use this occasion to try and teach us something. For example, during a favorite ride down a fairly steep hill in the Park, he would turn off the car engine at the top of the hill and let the car roll down. He would explain the mechanics behind the working of the clutch and gears, so that we could understand how the mass of the car moved it down the slope when no resistance was applied by the gear. We would be required to repeat the physics of the lesson during the next trip.

The fun part of this was waiting to see if he would remember to apply the brakes, since he got so involved in his explanations!

One of the most valuable lessons that my father taught us was the importance of time. He would often remind us that "time and tide wait for no man." He could not tolerate anything that he considered a waste of time. Television definitely came under this heading, so through most of our childhood, there was no TV in our home. However, during the first Indo-Pakistan conflict, he was persuaded to rent a television for a few weeks for news from home. I am embarrassed to say that we four children were quite happy when there was a second conflict a couple of years later and the TV was rented again! Now, of course, I am truly thankful for his wisdom because in the end, we benefited in many ways, not least through the treasure of books he left us and a love for reading.

It sometimes seemed that reading was second only to breathing to him. He read about every topic and his knowledge about all kinds of subjects was vast. In fact, I thought I would be able to hold my own by studying biochemistry, but he knew more about my PhD subject than I did! He loved to buy books, old and new. There was no room in our house, including the bathroom, that did not have bookshelves in it filled with books on every subject. My father would come home from traveling the globe with his suitcases splitting from the weight of the books he had bought, and my mother would worry about where to put them.

My father never took time off. Even when at home, he would spend hours engrossed in study in his room, some days only coming out at meal times. He had created his own very personal atmosphere of warmth tinged with an air of mystery in his room. The room temperature was kept extremely high, even in warmer months, and incense sticks (agarbati) would be lit sending spirals of heavily perfumed smoke in all directions. A cassette of recitation of the Holy Qur'an would often be playing softly in the background. Hence the Word of God was never far from his thoughts, a fact strongly reflected in his writings. The heavy velvet curtains would frequently be drawn closed against the noise of the street outside. From an early age, we all knew not to raise our voices or run around inside at these times; even the phone was taken off the hook if it rang once too often.

My father adhered strongly to the maxim: Early to bed, early to rise, makes a man healthy, wealthy and wise. He would go to bed around eight or nine o' clock in the evening, and arise a few hours later to work in the silent hours before dawn when his level of concentration and creativity would perhaps reach its peak, sustained by a thermos of hot, sweet tea and some snacks that we would place by his bedside before sleeping.

My father's hospitality was famous; he loved to invite people to our house, which he called the "best restaurant in the city"; a tribute to my mother's cooking. His guests included ministers, diplomats and foreign dignitaries, professors and quite often his students.

Of course, once the Center in Trieste was established, my father's time at home decreased considerably, and we would see him only for a few days each month. But no matter where he was, he always kept in touch by letter or phone. After I moved to the United States, my father always tried to visit me despite his busy schedule. Once he came to New York for one day only; for a meeting at the UN, flying in by Concorde. I was living 50 miles away on Long Island; so he made sure that a car was arranged to pick us up and take us to the hotel where he was staying. The last time he visited me was in 1988, after I had moved to Los Angeles, when he was invited to lecture at UCLA.

One of the greatest events for the family was the trip to Stockholm for the Nobel Prize ceremony in 1978. Naturally, we all wanted to attend the ceremony. However, when we got there, we were told that children were not allowed at the ceremony. At the time, my eldest son was almost two, and my sister's son was a year old. At my father's request, this long standing rule was bent and both children were allowed to attend. The next day, there were more pictures of the two grandchildren in the newspapers than of the Nobel Laureates, with the headline "Toddlers play at the feet of Nobel Laureates."

Although my father had received many honors before, he was quite nervous on the day of the ceremony. He had decided to wear Pakistani clothes to honor his country, but we could not get the turban wrapped to his satisfaction, and he began to get quite upset. Finally, after several attempts by various people, the day was saved by the Punjabi cook at the Pakistani Embassy, who was able to get it right!

It was a very proud moment for us all when my father delivered the Nobel Address that night and and the words of the Holy Qur'an were recited for the first time in the four hundred year old palace in Stockholm. He quoted one of his favorite verses:

> *No incongruity canst thou see in the creation of the Gracious God. Then look again. Seest thou any flaw? Aye, look again, and yet again, thy sight will only return unto thee confused and fatigued, having seen no incongruity.* (67: 4–5)

I am sure that I have taken enough of your time, but I cannot close without making a brief mention of my mother, who had been looking forward to this event, but passed away three months ago. While she was a very remarkable person in her own right, her contribution to my father's success was tremendous. Her faith, support and sacrifice allowed him to pursue his goals wherever his work took him. She would have been very moved by today's program.

I cannot close without mentioning the greatest quality that I remember in my father; that is his utter humility and faith in God. In spite of all the honors and successes he achieved, my father never showed any signs of pride. He never thought of himself as a great man, and he never forgot his humble beginnings. Most of all, he never forgot that all he achieved was due to the Favor and Grace of God.

Finally I would like to thank the College for holding this great event today, and allowing me to be part of it; it has been a great honor and I thank you all for your patience and attention.

RECOLLECTIONS OF MY GRANDFATHER

SAIF RAHMAN
(Grandson of Abdus Salam)

In the Name of Allah, the gracious and ever merciful

I would first like to thank Imperial College, Sir Richard Sykes, Mike Duff and all of the organizers of this special occasion, not only for the respect and honor shown toward my grandfather, but for also giving me the privilege to speak to you today.

I am here as a representative of all of Prof. Salam's grandchildren. As grandchild number 3 of 13, I was very fortunate to have been able to spend some time with him before his passing. However, living across the world from him in California, what little time I got to spend with him was very precious. In fact, it was not until he visited us in California that I realized that he was more than just my grandfather. In 1989 grandfather gave a lecture at UCLA in Los Angeles. I was only about 8 years old at the time, and although I couldn't tell you a word about his lecture, which I'm guessing was about physics, what I do remember so clearly was this packed lecture hall, with crowds and crowds of people who couldn't fit in the hall standing outside and watching the lecture on a projection. There were not hundreds, but literally thousands of people there intently watching and listening to my grandfather. Maybe I was a little slow, but that's when I really started to get interested in who this great man was.

And what a great man he was! I am so proud to be able to call the man we have been honouring today, my grandfather. We have heard earlier today from some of the respected speakers about the amazing scientific achievements and his accomplishments in the field of theoretical physics, and about the International Center in Trieste, Italy that he founded. But the legacy that I can truthfully say he has left to all of us and instilled in all of us was his desire and passion for serving humanity.

Although amongst the 13 of us grandchildren, we have gone into a variety of fields such as engineering, law, and medicine, and some are still studying in school, we all have this built in sense and drive to help those who are less fortunate.

If I can give a personal example, as a lawyer in America where I practice, it is very easy to become part of the money-making machine, filing lawsuit after lawsuit. But when I decided as a teenager that I wanted to study and practice law, grandfather gave me one piece of advice that I will always remember. He said Saif, I honestly don't care what field you decide to enter, as long as you can serve humanity and serve those in need. As a public defender, I am an attorney who is appointed to defend and help those who cannot afford an attorney at no cost. Although this is not the most glamorous, and especially not the most lucrative field of law, I can assure you that it is the most rewarding and gratifying.

As honoured and privileged as I am to be a grandson of Prof. Salam, I can also tell you that it was not the easiest thing. As you have heard from my mother before me, grandfather was very demanding from his children and grandchildren and only expected the highest of standards. It could be tough when you had a Nobel Laureate critiquing your homework. I remember one time, actually when he was in Los Angeles for the lecture I spoke of earlier, and he came to visit my school. I was in grade school at the time and I knew he was coming that day so I made sure I used my best penmanship. I really put in quite a bit of effort to get my work to look perfect. Grandfather went around the class examining the various papers we were working on and praising my other classmates for their penmanship. When he got around to my desk, I displayed my work to him, expecting some praise and compliments on my fine penmanship that I worked so hard on. Instead, he glanced at my paper, and turned away without saying a word. Later in the evening when I asked him what he thought of my work, he said that my penmanship was good, but don't expect to get praise for something that you are supposed to do well anyway.

What is interesting about that is grandfather actually lived his life with so much humility. For someone who had achieved such high rank and status among his peers, it is almost incomprehensible how humble he was. My grandmother used to tell us that when he would come home from a long overseas trip or even from Imperial college after a day of hard work, the first thing he did after entering the house was to make sure the driver was taken care of, given a meal or at least some tea. THIS was the behavior of a world-renowned scientist who had been in the presence of royalty and actually had had audiences with the Pope.

Another point that he made sure all of his children and grandchildren abided by was to respect our teachers. There was no blaming of professors or teachers for lousy marks in school. After he won the Nobel prize, he visited Pakistan and India. In India, he sought out and found his primary school teacher in the village of Jhang. His teacher was now at a very old age and could not even sit up. Grandfather put his Nobel prize on his teacher's chest and said "this is for you." The humility and respect that he showed towards others was unbelievable.

I would like to close my remarks with a personal observation that I have made regarding my grandfather and the way he lived his life. He could have devoted 100% of his time to his research in physics, however he never forgot where he came from and the adversities he faced coming from a Third World country. He paid so much attention to making sure other scientists from developing countries were given the means and opportunities to excel in their respective fields that I find myself even more proud of his humanitarian endeavours than his great scientific achievements.

I thank you for your time and I know my grandfather would be humbled, yet honoured by this special gathering.

Thank you.

MESSAGE FROM THE WORLDWIDE AHMADIYYA MUSLIM COMMUNITY

MIRZA MASROOR AHMAD
Head of the Worldwide Ahmadiyya Muslim Community

Abdus Salam was an Ahmadi Muslim from Pakistan, a renowned theoretical physicist who received the Nobel Prize in 1979 for his work in electroweak theory. Although he was the first Muslim Nobel Laureate, Pakistan's military dictator at that time could not admit that its brilliant scientist was a Muslim citizen. Dr Salam's entire award was devoted to the furtherance of education: he did not spend a penny on himself or his family.

Salam believed that scientific thought was the common heritage of mankind. He advocated that developing nations needed to help themselves and invest in their own scientists to boost development and fill the gap between the rich North and the poor South of the planet. Salam was the founder and Director of the International Centre for Theoretical Physics Trieste, Italy, and he also founded the Third World Academy of Sciences. He passionately promoted the advancement of science and technology.

Above all, Abdus Salam was a devout Ahmadi Muslim, whose religion was inseparable from his work and family life. He once wrote: "The Holy Quran enjoins us to reflect on the verities of Allah's created laws of nature; however, that our generation has been privileged to glimpse a part of His design is a bounty and a grace for which I render thanks with a humble heart."

Professor Dr Abdus Salam was a role model for the Ahmadiyya Muslim community and for all research workers. Many Ahmadi students today follow in his footsteps in an attempt to improve and advance in their respective fields.

We have a responsibility to enable all talent to scale to new heights. It would be grossly unfair to improve immigration, economic or social barriers to prevent such talent from wherever to blossom.

The Imperial College deserves to be complimented for giving birth to so many Nobel laureates. Offering scholarships and enabling students from poor countries to undertake their studies is indeed a matter of great pride for this institution. It deserves our gratitude and our best wishes.

PRESENTATION OF THE ABDUS SALAM IMPERIAL COLLEGE PHYSICS GRADUATE PRIZE

Dr. MIRZA AHMAD
President, Ahmadiyya Muslim Youth Association, UK

In the name of Allah, Most Gracious, Ever Merciful

Distinguished speakers and guests, members of Professor Abdus Salam's family, the students and alumni of Imperial College, peace and blessings of Allah be upon you.

First of all, let me congratulate you on the Centenary of Imperial College. We hope and pray that the College has an even more distinguished trajectory over the next 100 years and beyond.

Congratulations and thanks are also owed to the College, and the Physics department in particular, on hosting this event, which commemorates Professor Abdus Salam's 50th anniversary of his arrival at Imperial College.

I would like to take this opportunity to briefly introduce the Ahmadiyya Muslim Association and its Youth Association.

The Ahmadiyya Muslim Association was formed in 1889 in a small town called Qadian in India. It was founded by Hadhrat Mirza Ghulam Ahmad (may Allah bless him), who claimed to be the Reformer of the Age.

Now spread in over more than 175 countries across the world, and with a membership running into tens of millions, the core message of the Community is one of peace and understanding — indeed — the motto adopted by us is "**Love for All, Hatred for None**".

The Community has been led by successive Heads, referred to as a "Khalifa". Hadhrat Mirza Masroor Ahmad is the fifth successor to head the Community.

Professor Abdus Salam is a member of this Community.

The Ahmadiyya Muslim Youth Association was founded in 1938. Its key principles are of service to humanity, the community, the country, and of instilling in its members an ever lasting yearning for knowledge and education.

We firmly believe that a good education, coupled with personal development provides dividends both for the individual concerned, and for society as a whole. We have been undertaking these activities for more than 70 years. Our motto is that "**A Nation cannot be reformed without first reforming its Youth**".

We host a multitude of events annually. These vary from sports events, through to academic competitions, charity events, career fairs, forums and workshops.

A key focus is charity work. A variety of events are held all over the country every month, from hospital visits to visiting the elderly and infirm, voluntary services, and of course fundraising for National Charities.

Our volunteers have worked across the globe helping out where natural disasters have struck, from Kobe in Japan, to the Tsunami Relief Effort in Indonesia, and to Pakistan for the South East Earthquake.

Back at home we are actively involved with a multitude of national charities in providing volunteers for their respective events, such as providing marshalling for the Leukaemia Research Bikeathon's.

Our flagship event, a Charity Walk and Run, has been raising funds for more than 20 years. Our last Charity Challenge held in May 2007 is expected to raise more than £100,000 for UK charities. You are more than welcome to contribute towards it.

In order to promote our links with academic institutions across the country we have launched a number of initiatives and awards. The Abdus Salam Undergraduate and Graduate awards are part of this initiative and I have to thank Professor Duff and our own Student Affairs department for successfully bringing this to fruition.

This award is dedicated to Professor Abdus Salam, who for us embodied someone that we all want to emulate. In his search for decoupling the electroweak theory/ unified theory he perfectly coupled his search for knowledge, his thirst for learning, and his love for the Community and religion he belonged to. He himself once wrote:

> "The Holy Quran enjoins us to reflect on the verities of Allah's created laws of nature; however, that our generation has been privileged to glimpse a part of His design is a bounty and a grace for which I render thanks with a humble heart."

We actively encourage our youth to aspire to such lofty ambitions, our wish is that others will follow in the footsteps of this illustrious individual. The award being presented today will hopefully provide another catalyst to those aiming to achieve those goals.

Again, please accept my congratulations and thanks for hosting such a wonderful event and reception. Thank you.